# LIGHT: THE MYSTERY OF THE UNIVERSE

KHALIL SEYRAFI, Ph.D.
President, Electro-Optical Research Company

Lectures Presented in the
Department of Engineering and Science
University Extension
University of California, Los Angeles

1986
FIRST EDITION
A Publication of
Electro-Optical Research Company
Los Angeles, California

ELECTRO-OPTICAL RESEARCH COMPANY
2029 Century Park East, Suite 422
Los Angeles, CA 90067



Library of Congress Catalog Card Number: 85-80510
ISBN: 0-936581-00-X

Printed in the United States by
Pace Publication Arts Inc.
Inglewood, California

To Mahboob

# Preface

At its inception this book was intended simply to present a historical account of developments in the field of electro-optical research and, further, to give its readers some perspective on disciplines related to that field. As the writing progressed, however, the author began to realize that the story of past and current efforts to unlock the mysteries and explain the phenomena of the universe is of far more import than a mere recital of historical events. These efforts, furthermore, provide a basis for determining the development and history of optics. Consequently, *Light: The Mystery of the Universe* attempts simultaneously to explain the sequence of discoveries so important to the history of optics and the reasoning that brought order to them.

Optics is the science of light, which has played a decisive role in the natural evolution of plants and living organisms. It has inspired the disciplines of astronomy, photography, motion pictures, atomic physics, electronics, and computer sciences. Recently, through laser technology, space probes, and satellites, it has had a powerful impact on the field of communications. All these fields are inextricably woven into the fabric of present-day society. It is inconceivable that a human being can fully participate in twentieth-century life without at least a cursory understanding of these subjects. Without such an understanding, one is in peril of being lost in the mass of social, cultural, and scientific information. This book, then, outlines the developments in optics which have made the twentieth-century world what it is today.

Light has intrigued man throughout human history. Thousands of years ago his curiosity was aroused by the simple observation of heavenly bodies. As time went on, his curiosity led to research, experimentation, and discovery. In the same way, today's "Star Wars," laser and infrared sensors, and computers and robots are continuing to stimulate that curiosity.

*Light: The Mystery of the Universe* is the outgrowth of a course in electronics and optical systems I have taught at the University of California, Los Angeles, for the past twenty-three years. The book is the result of ongoing research regarding the modern world and the impact of light phenomena on human civilization. It presents the entire history of optics, dividing that history into four periods, and identifying the activities, main contributions, and eventual outcome of each. It does not, however, describe in detail each advance in optics, for that need has been filled by other writers, such as Whittaker, d'Abro, Jammer, and Born and Wolf. Rather, its purpose is to concentrate on the major contributions to the field, indicating how the study of light has contributed to human advancement, how research has progressed to its present level, and what direction it may conceivably take in the future.

Insofar as possible the material is presented in chronological order, though there are exceptions. For example, follow-up efforts after a particular discovery may not have reached fruition for several decades, and during that period there may have been other noteworthy developments. In such instances the chronological approach has been abandoned and discoveries are traced through their own evolutionary development. This procedure enables the reader to perceive the impact of any single idea on the overall progress in the field.

To facilitate reading, equations and formulas are presented minimally or omitted altogether. On occasion, however, the true meaning of scientific advances can be appreciated only by comparing key formulas. In the transformations of Galileo and Lorentz, Einstein's conversion of mass and energy, and the process of thermonuclear fusion, for example, nothing can replace the concise description of an equation.

The material here presented covers the scientific endeavors and the resulting achievements considered essential to an understanding of the complex world of today. It is the author's intention to show how, through many centuries, the study of light has been the source of innovation. After reviewing the progress, often slow and halting in earlier periods, which has led science to its present frontier in the twentieth century, the author summarizes the truly extraordinary accomplishments of scientists working in the fields of optics and electro-optics.

Ultimately, our multidimensional world is an endless mystery. The unknown has always challenged man's imagination and has led him to seek answers to questions about the universe. *Light: The Mystery of the Universe* attempts to fill some of the gaps in scientific knowledge by explaining developments that have come about through the study of light.

This book has benefited from the assistance of many good-natured people with whom I have had the opportunity to be associated. I would like particularly to thank my devoted secretaries Milrae Kruse and Beatrice Beggs for their tireless efforts in typing and retyping the manuscript. I am grateful to Chris Lansbury, Grace Stimson, and Jon Wilwerding for their editing of the manuscript, and to William Yeh for his assistance in researching several topics discussed in the book.

I would, furthermore, like to express appreciation to my students at UCLA who have encouraged me to write this book and who have made worthwhile contributions to its content. I am indebted, finally, to my sons, Sherwin and Shebly, and to my daughter, Carmen, for their constructive criticisms. To Shebly, in particular, I am indebted for his contributions to the discussion of quantum mechanics.

January 1986
Khalil Seyrafi

# Contents

# Introduction

Certain periods of history have been marked by notable scientific advances. The seminal accomplishments of the Greeks from 500 to 300 B.C., the creative achievements in the Renaissance, and the dramatic strides in the twentieth century are highlights in scientific development. These periods were all colored by the inquisitive spirit of the thinkers who sought to understand and explain the mysteries and phenomena of the universe. In other periods, such as the Middle Ages, scientific inquiry faltered, in part because of the inhibiting influence of church dogma.

Optics, the study of the genesis and propagation of light, is one of the most vital fields in science. In every era when science has flourished, optics has been in the forefront of scientific research. It has enjoyed a long and rich history throughout civilization because scientists and philosophers have been fascinated, as well as perplexed, by light and vision and by their interplay in the universe. Of all natural phenomena, light has attracted the closest attention in practically all fields of study and thought. Earthbound man's curiosity about what he could see in the sky has led him on a never-ending quest for knowledge of the universe, and that quest has always been initiated or furthered by the science of optics.

The evolution of optics through the ages has been impressive. From the mythical beliefs of the ancient Sumerians and Greeks to the discovery of the cosmos in the early seventeenth century, from the determination of the temperature and the composition of distant stars in the nineteenth century to the unlocking of the mystery of the atom in the twentieth, optics has influenced science, religion, and philosophy.

According to those early mythical beliefs, the cerulean dome of the sky was merely a playground for gods and demons. On the contrary, the modern "big bang" theory, attempting to explain the boundless universe, holds that everything exists on a cosmic scale, that we live on a tiny planet lit by a dying star in the midst of billions of other stars and galaxies. Thus we can comprehend only a small part of universal space, ranging from mere fractions of millimeters to hundreds of thousands of kilometers. Beyond a certain point we lose our sense of spatial dimension, as vast distances are represented simply by numbers in equations. The distance between the earth and the moon (250,000 miles), although minimal in comparison with the distance betwen the earth and the sun (90 million miles), seems to be much the same as we look at the two heavenly bodies. Advanced optical instrumentation allows the eye to see beyond its natural potential, giving us new visual experiences that, along with the contributions of computers, have substantially increased the capacity of the human mind. But just as the human eye has physical limitations, so the mind is unable to grasp the dimensions of the universe, estimated in trillions of trillions of meters ($10^{24}$ m), or atomic dimensions, estimated in trillionths of a millimeter ($10^{-15}$m).

Today we live in three worlds that are distinct from one another:

1. *The world of ordinary physical objects visible to us in our everyday life.* It is the world of common sense. Our actions are based on our perception of this world. It comprises our interactions with our surroundings and with other people, which are guided by our instinct for survival. In this comprehensible world our history and our traditions are written and formed. In it, time is measured in seconds and space in meters. The natural physical forces recognizable in it are confined to gravity and electromagnetism.
2. *The world of the atom.* Here time is measured in billionths

of milliseconds (1 femtosecond = $10^{-15}$ s) and space is measured in femtometers (1 femtometer = $10^{-15}$m). The atomic world cannot be comprehended by common sense. Although its manifestation in terms of energy is common knowledge, thorough comprehension of it requires an in-depth understanding of mathematics and physics. Thus the atomic world, unlike the ordinary world, is comprehensible only to those who possess a knowledge of quantum mechanics. In the atomic world, three forces are at work: electromagnetism, the weak nuclear force, and the strong nuclear force.

3. *The world of the universe.* Still largely unknown to us, the universe is a mystery that seems to grow rather than to diminish with each new discovery. On the cosmic scale of the universe, time is measured in billions of years and space is measured in light-years or in trillions of miles. Such temporal and physical dimensions render this world incomprehensible to us. The reference points are the sun, the planets in our solar system, and clusters of distant stars visible to the naked eye or seen with the aid of telescopes. From beyond these visible elements of the universe we receive and record signals of rays emanating from stars and galaxies that are even farther away. These signals suggest that the universe is still expanding, an idea postulated by the big bang theory. Physical events in the universe are the outcomes of the four forces of gravity, electromagnetism, and the weak and strong nuclear forces. This third world is one we can see only as it was, never as it is. Light from distant parts of the universe must travel for 12 billion years before it reaches the earth. In other words, the images brought to us by this long-journeying light will be 12 billion years old. They, and they alone, tell us what was happening in the universe an uni-

maginably long time ago.

The science of optics describes the processes by which natural phenomena in these three worlds are probed and explained. Its historical discoveries show how man's knowledge has increased through the ages and how some of the mysteries of the universe have been resolved.

Light has been a pivotal element in all three worlds. Beyond its obvious function of illuminating the ordinary world, it has been the focus of numerous experiments in all three worlds. Through such experiments, great scientists in all periods of history have played a role in the evolution of optics, the science of light. They have proposed, and are still proposing, theories to explain the properties and the functioning of light, often before conducting experiments to validate the theories. So close has been the association between great minds and the development of optics that one might ask whether geniuses are attracted to the phenomenon of light and its physics or whether light itself penetrates as a flash of genius into the minds of its investigators.

Early civilizations based their explanations of visual phenomena in the heavens on superstition, imagination, and theoretical beliefs. Later, as man's power to reason evolved, his ideas about the universe relied more on common sense. Still later, theories unsupported by experimental evidence yielded to those based on Galileo's discovery of the telescope, which made possible a better comprehension of the cosmos, on Newton's explanation of the particle nature of light, and on Huygens's and Fresnel's discussions of the wave nature of light. These advances gradually removed optical knowledge from the reach of common man and placed it in the hands of scholars. Through the discovery of and continuous improvements in the telescope, recently augmented by space probes, optics has unveiled some of the mysteries of the cosmos. Larger and more powerful telescopes, still in the design stage, will probe farther into the cosmos and will enable man to see 12 billion years into the past.

The merging of optics with electricity and magnetism, as an accidental result of nineteenth-century experiments, set the stage for Einstein to formulate the theory of relativity. In the early twentieth century the velocity of light was established as the limiting speed of any matter in the universe. Before mid-century, intensive measurement of the optical spectra of natural elements had opened the atomic world to exploration. Perhaps the most spectacular contribution of optics is attributable to this opening up of the atomic world: the development of the atomic theory in the early twentieth century. The optical signatures of basic elements, such as hydrogen and helium, were used in continuing efforts to confirm or modify the new theory. Through the integration of optics with electronics and their application in space probes, optics has played an important role in the achievement of outstanding scientific advances.

The advent of photography in the early nineteenth century and of motion pictures in the latter part of that century, both related to optics, transformed the social and cultural habits of mankind. The merging of optics and electricity in the twentieth century, giving birth to the field of electro-optics, led to the development of television. Following upon a long series of unrelated innovations beginning in 1820, the invention of television in the mid-1920s not only enhanced motion picture production but created an entirely new medium of entertainment and education. Old customs, such as gatherings in the town square where people exchanged news and local gossip, yielded to modern ones centered on TV sets. Today, television is an inseparable component of everyday life for most people in the world.

Ironically, the merging of optics with electricity and magnetism (from the work of Faraday and Maxwell), the discovery of the photoelectric effect (by Einstein), and, finally, the association of optics with atomic particles manifested by quantum mechanics revealed light itself to be more mysterious than ever.

Although science knows no boundaries, either geographical

or chronological, history has proven that major scientific advances tend to occur in and shift with the power centers of the world. This pattern is witnessed by the interplay between power and scientific innovation in ancient Greece; in the Middle East during the Middle Ages; in western Europe during the Renaissance, the industrial revolution, and the nineteenth century; and in the United States in the twentieth century. Thus historical progress in optics may be traced through four stages:

1. *Prescientific period* (3000 B.C. to seventeenth century): Marked by slow progress in optics, the period ended with the development of the law of geometrical optics. Scientific advancement from ancient times through the medieval era was guided by a desire to achieve a better understanding of the function of human vision. Prominent in the period were Democritus, Empedocles, Ptolemy, Plato, Aristotle, Euclid, Alhazen, Bacon, da Vinci, Kepler, Snell, and Descartes.
2. *Early scientific period* (seventeenth century to early twentieth century): Scientific boundaries were extended by the invention of the telescope and the microscope, the formulation of classical physics, and development of the radical wave theory of electromagnetism. The discovery of X rays and radioactivity opened new avenues in scientific thought at the turn of the century. The works of Galileo, Newton, Huygens, Fresnel, Young, Kirchhoff, Maxwell, Röntgen and Becquerel were important in this period.
3. *First half of the twentieth century* (1900-1950): Beginning with the revolutionary theories of the atom and the special theory of relativity, this period also saw the introduction of the quantum theory. It concluded with a surge of new technology after discovery of the transistor and the computer. Comprising the works of Planck, Einstein, Curies, Hubble,

de Broglie, Pauli, Heisenberg, Schrödinger, Dirac, and Fermi, it was also marked by the shift of scientific achievements from Europe to the United States. By the second quarter of the century the United States had replaced European countries as the center of research in optics.

4. *Second half of the twentieth century* (1950 to date): Advances in optics have been truly remarkable. The discovery of the laser, an extremely powerful light source, made possible revolutionary applications to numerous problems, among them some of the oldest dilemmas in optics. Other significant discoveries were holography, optical detectors, solid-state devices, computers, and the integration of optics with a mosaic detector to fabricate optical sensors, thus giving rise to the development and application of electro-optical systems. These sensors, mounted on space probes, transmit space-to-earth data, supplying an abundance of new information about other planets in our solar system as well as about more distant heavenly bodies. Thus this period has witnessed the birth of space astronomy. Previous advances in radar technology have contributed to further progress in electro-optical systems. The work of Feynman, Schwinger, Gabor, Townes, Maiman, Weinberg, Salam, Glashow, and Gell-Mann are among the outstanding scientists of this period.

# 1

# Dominance of the Myth: The Prescientific Period

## 1.1 Introduction

From the beginning of civilization thousands of years ago to the early seventeenth century progress in optics and in the understanding of light was very slow. During this prescientific period, beginning with ancient man, extending through the era of mythical beliefs, and ending with the discovery of the telescope, the human eye was the only optical device to explore and to be explored. The meager discoveries of the period were limited to evaluations and assessments of visual perception of the environment. Light was considered the source of and the reason for vision. Seen as a simple element related to fire, one of the four elements believed to compose the earth, it was observed to be abundant during the day and virtually nonexistent at night. Man's activities were performed during the diurnal period of light, and his rest period coincided with the absence of light.

Progress in optics was therefore limited primarily to intellectual views on and the understanding of human vision. What eventually became the visual ray theory dominated optics during the prescientific period. According to this theory, the ability to visualize an object depended on a visual ray that emanated from the eye and extended to reach and touch the object. Although the visual ray theory was proven false almost 300 years ago, it is amazing that man accepted it as truth for more than two thousand years. Competing theories occasionally did arise, but for various reasons none of these were able to gain ascendency over the visual ray theory. One such reason was that, during the early part of this period, some distinguished philosophers actually discouraged experiment, maintaining that truth is revealed by the mind and not by the senses. Though it seems foolish in light of modern scientific progress, this view unfortunately did influence the methodology of the era. Later the rise of church dogma would have an influence similarly hostile to the overthrow of the visual ray theory and to scientific advance generally.

Man's knowledge of the earth and the heavens was based on his mythical beliefs, which were influenced by leaders of the church and by such famous philosophers as Plato and Aristotle. Once a philosopher had established himself as proficient in a specific subject, his convictions on other subjects were instantly accepted as truth. Thus his opinions about a new and possible threatening theory conceived by another mind were a controlling factor in the destiny of the new theory. This association of power with prestige stifled the creativity of scientific and philosophical minds from early civilization up to the seventeenth century.

Lack of significant experimental evidence forced man to subscribe to the leading philosophical dogmas and the mythical convictions of this ancient world. Because of the instrinsic relationship between optics and vision, initial advances in optics were spurred primarily by the intriguing exploration of vision. The study of optics began with playing with light and offering imaginative explanations, a process now known as experimenting and hypothesizing.

The oldest record of man's view of the universe was that of the Sumerians in the Middle East (3000 B.C.). They believed the earth was the center of the universe, that it was flat and fixed in space. The thin dome covering it was the heavens, with nothing beyond. The heavens, containing the sun, the moon, five planets, and the Milky Way, were the home of gods and demons who traveled among them. The five planets were Saturn, Jupiter, Mars, Venus, and Mercury. The inexplicable incidents affecting man's life on earth, otherwise known as natural phenomena, were the results of warring between the gods and the demons in the heavens.

## 1.2 Vision of Ancient Man: Era of Speculation

The theories of light and vision have strongly influenced the development of science throughout civilization. Many scientific findings and the developments leading from them have their basis in and were nurtured by discoveries in optics. Study of optics began around 2000 B.C. The earliest optical invention, prompted by man's desire to see his own image, was the mirror. Bronze or copper mirrors were made as early as 2000 B.C. Recent discoveries in the pyramid of Sesostris II (1900 B.C.) in the Nile Valley included a relatively perfect mirror.[1] And in a provocative and imaginative story, Exodus (1200 B.C.) recounts Satan's recasting of a woman's looking glass into a "brass laver." But the invention of the mirror did not play a central role in the first genuine advances in thought about optics. What figured centrally in these advances was instead the intriguing nature of vision. It was the intrinsic connection between the study of light and the study of the eye that caused optics to take its first halting steps forward.

*Democritus (c. 460 - c. 370 B.C.)*

The theories on the nature of light formulated by the Greek philosophers Pythagoras, Democritus, Empedocles, Plato, Aristotle, and Euclid[2,3,4] among others, were based upon their understanding of human vision rather than on analysis of the physical phenomenon of light. Two of these theories, the "atomic ray" of Democritus and the "visual ray" of Empedocles[5], prevailed until the Middle Ages.

*Empedocles (c. 490-430 B.C.)*

In the fifth century B.C., Democritus (460-370 B.C.), a pioneer in atomic theory, proposed an explanation for man's vision. He believed that the universe is filled with an infinite number of invisible particles or atoms* and that physical phenomena are produced by the interaction of atoms swarming in empty space. Man's visual perception, he claimed, was an optical illusion, as there existed only this interaction between atoms. Vision actually occurred only when an object shed its atoms into space. These atoms were received by the eye, then intercepted by the soul; thus visual imagery was perceived. Democritus's atomic theory of vision is a crude outline of what is now our best explanation of how we see things. That a man of the fifth century B.C. could anticipate the accepted modern theory of vision should command wonder and respect in the twentieth century A.D.

*The term "atom" comes from the Greek word "tomos," meaning to cut; therefore, "atom" means that no further subdivision can be made.

*Plato (427-347 B.C.)*

Parallel to Democritus's atomic theory was the visual ray theory of the Greek scientist Empedocles (490-430 B.C.), who believed that an object was perceived when light radiating from the eye touched it. According to Empedocles, all matter in the world was composed of four elements: earth, air, water, and fire. The pupil of the eye radiates a visual "fire" just as a burning lantern emits light. One might trace the origin of Empedocles' visual ray theory to the concept of perception. The hand extends to reach, touch, and thereby to sense an object. Similarly, the eye emits a visual ray to reach, touch, and sense an object, thus creating vision.

Plato (427-347 B.C.), the celebrated Greek philosopher, also subscribed to the visual ray theory. He was so strongly opposed to Democritus's atomic theory that, according to one account, he confiscated the latter's manuscript and burned it. Plato, however, modified the visual ray theory. He proposed that rays emanating from the eyes united with sunlight, and that vision was produced when visual rays intercepted rays emitted by the object. In his Dialogues[6], Plato discussed the bending image of an object immersed in water. He suggested that color was produced by the transfusion of particles of visual rays and object rays. Particles of the same size created a transparency but particles of different sizes yielded color. Ironically, Plato cautioned against testing the visual ray theory by experiment, as he mistrusted physical realities.

Plato, a student of Socrates and teacher of Aristotle, was the second of the three great philosophers of ancient Greece. His chief work is the famous Dialogues, which contains discussions of ethics, politics, and metaphysics. It would be impossible, in a few sentences, to describe the historical influence of Plato's thought on these subjects. What can perhaps be said with confidence is that Plato is still read seriously, at least by students of philosophy and political theory.

One belief of Plato's that is relevant here is his belief that the visual ray theory was not to be tested by experiment. Experiments require us to use our senses in order to observe experimental results. But it was Plato's view that using our senses to observe the world is a wrong-headed way to go about confirming theories of nature. It is perhaps hard to say how much influence this very foreign view exerted. What is clear is that Plato's advocacy of the visual ray theory and his reputation as a profound thinker helped to keep the theory alive until the Middle Ages.

An institutional accomplishment of Plato's, in addition to his intellectual contributions, was his founding of the Academy as an institute for systematic pursuit of philosophy and science. Some regard the Academy as history's first university.

In 350 B.C. Aristotle (384-322 B.C.), Plato's pupil, rejected his teacher's visual ray theory on two grounds. First, he argued that it was not reasonable to assume that visual rays could travel as far as the stars. Second, he reasoned that if visual rays truly originate in the eye, one should be able to see in the dark without the aid of any additional light source. Aristotle's own idea of vision was that the space between an object and the eye was an arena in which the projection of an image traveled from the object to the eye. This idea is comparable to the modern concept of optical transmission through the atmosphere. By receiving the image of an object projected through space, the eye perceives the existence of the object.

As a youth, Aristotle was sent to Plato's Academy, where he studied for twenty years. He would later serve as tutor to Alexander the Great and found his own philosophical school, the Lyceum. More important, he became, arguably, the single greatest thinker of ancient culture. Even less than in the case of Plato is it possible in a few sentences to describe Aristotle's historical influence. He authored a theory of physics which would not be overthrown until the time of Galileo, twenty centuries later. He fathered the science of biology and the systematic study of formal logic. His metaphysics provided much of the conceptual framework within which St. Thomas Aquinas developed a theology still important to Roman Catholicism. And modern moral philosophy, in England and America, still concerns itself with developing Aristotle's ideas, more than two millennia after his death. It is simply not possible to name another thinker with a record like Aristotle's.

*Aristotle (384-322 B.C.)*

The theory of vision continued to evolve when the mathematician Euclid (c. 300 B.C.) incorporated the visual ray theory into his own theory, presented in *Optica.* He argued that distant objects could not be seen because they fell behind interfering rays emanating from closer objects, thus obscuring vision. Euclid's support, along with Plato's, helped to sustain the visual ray theory until the Middle Ages.

After studying the reflection of light, Euclid made several interesting observations. His experiments with plane and curved mirrors led to two major discoveries:

1. Light travels in a straight line.

2. Incident and reflected rays create equal angles with a mirror's surface (the law of reflection).

Euclid also observed that rays reflected from concave mirrors come to a focus, thus explaining why a concentration of sunlight is powerful enough to cause combustion. In this "burning glass" phenomenon, when a convex glass is positioned so as to focus sunlight on paper, leaves, or other easily ignited material, a pinpoint of intense radiation is created and the heat produced sets the material on fire.

*Euclid (c. 300 B.C.)*

Euclid is also known for his contributions to geometry. In Euclidean coordinates, a straight line is the shortest distance between two points (later shown not always to be true); parallel lines never meet; and the sum of the angles of a triangle is 180°.*

The Roman philosopher Seneca (3 B.C.-65 A.D.) was intrigued

*An example of non-Euclidean geometry is that of the surface of a sphere.

by the remarkable magnification property of a glass globe filled with water. Earlier Greek scientists had also observed the bending of light rays at the interface between two media, better known as refraction. Both Cleomedes (50 A.D.) and Claudius Ptolemy[7] (130 A.D.) studied this phenomenon. Noting the apparent displacement of objects when viewed through water, Ptolemy observed that the ratio of the angles of incidence and refraction rays with the normal to the interface remains constant for all directions of incidence. He also discussed the refraction of light from celestial bodies through the atmosphere at various altitudes. In 150 A.D. Hero asserted that light rays reflected from a mirror would always follow the shortest possible path.

*Ptolemy (c. 130 A.D.)*

The Romans made relatively few advances in optics, except for the second-century physician Galen, who raised two objections to the atomic ray theory of Democritus. First, he argued that the images of large objects such as mountains could not be squeezed through the tiny pupil of the human eye. Second, he held that an object cannot emit enough radiation to reach all human eyes that might simultaneously be observing it. In propounding his own theory of vision Galen assumed that optic nerves are a conduit of visual spirits or "pneuma" originating in the brain. These spirits agitate the air near the eyes. By contacting objects, the air produces sight, becoming an extension of the eyes. But Galen's pneuma theory was overshadowed by the theories of Plato and Aristotle, and his contribution to optics went unrecognized until the time of Leonardo da Vinci.

## 1.3 Light in the Dark Ages: Stagnation in Science

Western science remained dormant for centuries after the fall of Rome in 475 A.D. Religious obstacles prevented the study of optics and the fabrication of instruments designed to aid the human eye. Tampering with light as a mere physical phenomenon was just as heretical as inquiring into the chemistry of the Eucharist. The general belief was that mirrors, prisms, and lenses reflected visual misrepresentations of reality, not reality itself. Man-made implements for enlarging, reducing, deflecting, or inverting visual images were believed to distort the truth. Devout Christians and renowned philosophers of the period would have nothing to do with what they regarded as trickery.

*Abu-Ali Sina (Avicenna) (980-1037)*

During the Golden Age of Islamic rule, from the seventh to the thirteenth century, the Middle East became the intellectual center of the world. As this period began, science was very much as the

Greeks had left it. Persian and Arab philosophers were able, however, to use Greek accomplishments as a starting point for advancing scientific ideas. Their accomplishments include construction of the first astrolabe, for measuring the altitude of stars, and the discovery of chemical substances useful in medical applications. By the close of the period, Arab and Persian thinkers had advanced science far beyond the level reached by their European predecessors, though, from the ninth through the eleventh century, several of them revived the Greeks' work in the study of human vision. One of the most famous Arab philosophers was the illustrious Yaguh-ibn-ishag-al-Kindi[8] (813-870). Al-Kindi wrote many books, including one on optics in which he accepted the visual ray theory and defended Empedocles' idea. On the other hand, Abu-Ali Sina (Avicenna), a Persian philosopher who subscribed to Aristotle's view, rejected the visual ray theory, arguing that rays could not possibly reach the stars. He also rejected Galen's theory that eyes could agitate the air around them. Ibn-Sina is also credited with writing a basic text in medicine which was read by Western students of medicine for the next 700 years.

The greatest Islamic contributor to optics was the eleventh-century scientist known as Alhazen[9], who studied the functioning of optics have been documented in his comprehensive book, *Opticae Thesaurus Alhazeni libri vii,* translated into Latin in 1270. Alhazen was, after the eye and wrote seventeen papers on optics. Alhazen was, after Aristotle, the first to attack the visual ray theory on scientific grounds. He argued that one who stared at the sun felt pain; and that when one looked away from an intensely bright object, an afterimage persisted. These reflections brought him to believe that objects were the source of radiation.

Alhazen proposed that visual perception takes place when the light rays of any illuminated object reach the cornea of the eye and are transmitted to the eye lens,* where the image is then perceived.

*The word "lens" comes from the Italian word "lente" from "lentil" or "lente di vetro," meaning glass lentil.

His proposition was remarkable. In spite of the enduring influence of ancient Greek theories of vision, Alhazen presented original ideas in optics and supported them with proof. His proposition rejected the still influential ancient Greek theories of vision and replaced them with an original idea. In his studies of curved mirrors, Alhazen was the first to note the existence of spherical aberration,* an optical defect. This apparent defect led him to believe that a perfect focus using spherical surfaces is impossible because the extension of the object above and below the optical axis generates an out-of-focus image.

*Roger Bacon (1214-1294)*

During the twelfth century progress in optics was marked by the invention of eyeglasses, which were used initially to correct farsightedness. At the time, however, scholars were unable to explain the function of the curved glasses. They simply did not possess a sound enough knowledge of optical transmission. This knowledge would not be forthcoming until the discovery of the law of refraction which accompanied the invention of the telescope in the early seventeenth century.

By the thirteenth century, creative achievements in Islamic

*Rays of different locations on aperture generally do not focus at the same point at the focal plane of an optical device. The error associated with this phenomenon is called spherical aberration.

society had subsided. Baghdad fell to Mongolian invaders in 1258, and concurrently Europe began to experience an awakening of creativity. In the field of optics, however, advances were negligible, as the framework within which new concepts were formulated remained the visual ray theory, first enunciated by Empedocles and later modified by other philosophers.

In the first half of the thirteenth century, Robert Grosseteste (1175-1253), an Englishman, tried to reintroduce Greek scientific knowledge in the West. He, too, was an advocate of the visual ray theory. A younger contemporary, the English philosopher Roger Bacon,[10] made significant contributions to science, believing that scientific knowledge would fortify Christendom against the threat of Moslems, Tartars, and even the Antichrist. He agreed with Alhazen's view that visual perception of an object is caused by the light shed by the object. But Bacon, following the visual ray theory, argued that ordinary objects did not produce optical rays and therefore could not produce vision. These objects needed visual rays to complete visual perception.

By investigating the scientific aspects of lenses and mirrors, Bacon evaluated the brightness of images of various objects. By combining lenses, he foresaw the possibility of the construction of a telescope. When Bacon presented the results of his study in court, describing the operation of lenses and mirrors and the nature of rainbows, he was accused of having used black magic and witchcraft. Although he did most of his scientific writing in secret, Bacon was imprisoned and believed to be in league with the devil and at odds with Pope Clement IV.[2]

Another thirteenth-century Englishman, John Pecham, presented the same basic theories as Bacon in his book on optics, *Perspective Communes.* Pecham's book became the most popular medieval treatise on optics, going through eight printings during the sixteenth century.

After the scientifically weak fourteenth century, the fifteenth century gave birth to a genius, Leonardo da Vinci[11] (1452-1519). He was to infer from the similarity between sound echoes and optical reflections, and was the first to state, that light might be a sort of wave. First analyzing wave motion in water, da Vinci moved on from water to sound and then to light:

> *Just as a stone thrown into water becomes the center and cause of various circles, sound spreads in circles in the air. Thus every body placed in the luminous air spreads out in circles and fills the surrounding space with infinite likenesses of itself and appears all in all and all in every part.*

Da Vinci deduced that the eye operated as a camera obscura, which consisted of a pinhole in a wall admitting light into a darkened room; an inverted image of the outside scene appeared on the opposite wall. For centuries, this device was used to observe solar eclipses.

At first, da Vinci defended the visual ray theory, referring to the ability of animals to stalk and kill their prey in darkness. In later writings, however, he reversed his position. He rejected the visual ray theory by reasoning that if the eye itself projected a visual ray, distant objects could not be seen at the instant the eye opened. Not knowing the velocity of light, he speculated that a visual ray took more than a month to reach the sun.

Best known for his artistic talents, da Vinci should be recognized also for the universality of his spirit. His ability to create graphically precise representations and his unusual powers of observation were revealed equally in his science and in his art. He remains an almost unique example of a human being who combined artistic and scientific genius. It is in the synthesis of art and science that da Vinci, in his genius, developed his own unique "theory of knowledge," in which the visual organ was the greatest gift to man.

In the latter part of the sixteenth century, the Italian physicist

Giambattista della Porta (1535-1615) elaborated further on da Vinci's camera obscura. Della Porta inserted lenses in the pinhole of the da Vinci camera obscura and thus improved the quality of its image. He also discussed multiple mirrors and combinations of positive and negative lenses in his *Magia Naturalis* (1589).

*Leonardo da Vinci (1452-1519)*

Initially explanations of human vision were entirely speculative. Centuries passed before the human eye, light, and the cosmos were studied systematically. It was not until the seventeenth century that man began a consistent and truly scientific method of stating hypotheses, making them mathematically precise, and proving or disproving them through controlled experimentation. The road to the scientific study of optics was a long one, for thinkers were concerned more with the desire to understand human vision than with systematic inquiry. Things began to change in the seventeenth century, however, as modern scientific method was brought to bear on the phenomenon of light.

# 2

# Lighting Science and Then the World: The Early Scientific Period

## 2.1 Introduction

The period from the seventeenth through the nineteenth century was marked by tumultuous scientific activity. Optical innovations led to significant advances in physical science and astronomy and helped to restructure man's social and cultural behavior. The period began with the invention of the telescope, which revealed new truths about the cosmos and negated ancient beliefs. Many phenomena formerly seen as mysteries were explained during this time. Oddly enough, toward the end of the period, deciphering of the new discoveries only confounded scientists by revealing the existence of still deeper mysteries.

A distinguishing characteristic of this period was its novel approach to establishing scientific facts by experimentation, application of mathematics, and formulation of observed results. Applied to optics, these new methods revealed that light had extraordinary properties never before realized. For example, it became clear that color composition was a capricious expression of light, as exemplified by the magnificent color spectrum produced when white light was passed through prisms and glasses (refraction), the systematic shading of its components from red through ultraviolet, the beautiful reflection of light by raindrops, and the rainbow. Such phenomena were enthusiastically investigated and analyzed throughout the period.

Two new theories of light were developed: the particle theory and the wave theory, which competed fiercely with each other. Both theories helped to remove the 2,000-year-old visual ray theory from further consideration.

Although the particle theory of light dominated the latter part of the seventeenth century and the entire eighteenth century, the wave theory started to gather momentum only during the early nineteenth century. By the second half of that century, it had been generally accepted by the scientific community as the correct theory.

The golden age of electricity and magnetism was concurrent with the development of optics. Experimental investigations from 1775 to 1865 in electricity and magnetism brought about significant discoveries, but it was not until 1850 that a crucial relationship between electricity and magnetism was discovered. This relationship, together with experimental results in optics, led to a turning point in science: the significant discovery that light is an electromagnetic wave. The period concluded with disclosure of the celebrated classical equations of electromagnetism. Shortly afterward, radio waves were generated and tested, confirming the wave nature of electromagnetism.

In the nineteenth century it was discovered that each natural element expresses itself distinctively on the optical spectrum. By applying this technique in determining the optical signatures of the elements on earth, scientists were able to establish the composition and temperature of stars. Once techniques for measuring the speed of light had been developed and their accuracy ascertained, a snowball effect followed.

The nineteenth century also witnessed the creation of the amazing world of photography and motion pictures. Photography started the process by recording and preserving a fixed or a moving permanent optical signature of human and other objects. With the perfection of techniques for the camera, phenomena invisible to the naked eye — such as the trajectory of a bullet, the movement of splashing water droplets, or the motion of a flower's petals as it starts to bloom — became visible. Furthermore, the invention of the microscope opened the new world of microorganisms, hitherto undetectable by man.

*Galileo Galilei (1564-1642)*

Today these new fields enrich our daily lives, especially in education and entertainment. Photography and motion pictures have completely transformed society and have, indeed, set a new mood for people. New emotional interplay has been created by the power of the camera and by its application to optical devices. Reminiscing with photographs (not stories), seeing a filmstrip of a foreign land (not reading about it), and fearing war from seeing its results (not just hearing about them) are minor examples of the role that photography and film play in our society. They intensify human moods —of pleasure, fear, affection — as well as enhancing entertainment, education, and politics.

## 2.2 Shaking the Unshakable: Progress in Optics

The period from the seventeenth century to the beginning of the twentieth century was one of the most flourishing in the history of science. It marked the beginning of a scientific approach to the field of optics, which led to the invention of the telescope and the microscope and hence to the discovery of other planets and their moons* and the discovery of microorganisms. The period was also marked by the confrontation of scientific findings with church dogma.

### 2.2.1 Probing the Cosmos: The Invention of the Telescope

It is unclear as to who actually invented the telescope, but records in the archives at The Hague show that on October 1, 1608, Hans Lippershew[2] (1587-1619), a Dutchman, applied for a patent on the device. Galileo Galilei[12] (1564-1642), an Italian mathematician, astronomer, and physicist, learned about the patent, and about 1610 he constructed the first astronomical telescope with a magnification power of 32. Through his telescope, Galileo, in a series of discoveries, found Jupiter's moons and Saturn's rings.

Telescopic observations of heavenly bodies had a profound impact on the reigning ideas of the time. The basis of the Aristotelian-Ptolemaic earth-centered universe was shaken. Only a few decades earlier the Polish mathematician Nicolaus Copernicus (1473-1543) had developed the revolutionary concept that the sun, not the earth, is the center of the solar system. His treatise, *De revolutionibus,* probably completed by 1530 but not published until 1543, when Copernicus lay dying, generated the first major battle between the church and science, although his revolutionary ideas remained dormant until Galileo's time.

Through his telescope Galileo observed that the moon's surface is not smooth but irregular, that the Milky Way is a mass of countless stars gathered in clusters, and that Jupiter has four satellites or moons. He found that the shape and the form of these celestial bodies are similar to those of earth. He also observed the phases of Venus, and he found "spots" on the sun. Galileo's detection of sunspots helped him to prove that the earth revolves around the sun. His observations confirmed the Copernican model of the universe, which

*See the discussion of astronomy in section 4.12.

holds that the earth is not unique but that it and the moon, Jupiter and its four satellites, and the other planets all revolve around the sun.

Galileo's development of the telescope and the observations it enabled him to make marked a turning point in science. His discoveries and his understanding of the solar system and the universe beyond it led to the emergence of the new science of cosmology. Today the telescope, enhanced by extensions of its initial capabilities, is the essential instrument for further investigation of the universe.

*Nicolaus Copernicus (1473-1543)*

Because of the intellectual restraints imposed by the Catholic church, European society did not embrace Galileo's new findings with enthusiasm. Most of the people in influential positions did not trust the telescope's revelations. They believed that the instrument was deceiving them and that what they were observing through it was an illusion produced by trickery. To be certain that he himself was not being deceived, Galileo frequently observed at close range the earthly objects he had viewed through his telescope, to verify their physical existence.

Galileo was persecuted by the papacy from 1616 until 1633, at which time he was forced publicly to denounce the results of his findings. He was ignominiously confined to his home in Arcetri, near Florence, until his death in 1642. Galileo became blind toward the end of his life. Speculation is that the damage to his eyes resulted from his telescopic observations of the sun.

Galileo, in fact, created the modern conception of experimentation. In 1581, while in his first year at the University of Pisa, Galileo noticed that a swinging lamp always took the same amount of time to swing back and forth, no matter how wide the swing of oscillation. Later in life he verified this law of pendulum motion experimentally, suggesting that it could be applied to the regulation of a clock. By observing objects thrown into the air, he also demonstrated the theory of parabolic trajectory. Galileo's observations regarding gravitation and falling bodies later led Newton to formulate his second law of motion.

One of Galileo's most important experiments was to release a ball on an inclined plane and observe its acceleration. This experiment confirmed the law of uniformly accelerating motion, that is, that bodies on the same plane accelerate or decelerate uniformly. By still further experiments with falling objects Galileo proved as well that the rate at which a body falls does not vary with its weight. This put to rest the long-standing Aristotelian error that bodies of different weights fall at different speeds.

It was a great achievement of Galileo's to establish these results experimentally, but it was a still greater accomplishment to provide a set of general laws, a theory of the motion of bodies which explained his experimental observations. Galileo occupies a unique place in the history of physics. For twenty centuries humans had thought about physics more or less in the way that Aristotle had. Galileo's work overthrew this paradigm. Not merely did he formulate and prove some of the most profound laws of physical nature; he was, rather, the first to do experimental science as we now think of it. This, together with his foundational work in gravitation and motion, makes him the father of modern mechanics and modern experimental physics.

Galileo's most celebrated equation is his coordinate transformation. If an object has space-time coordinates x, y, z, t in one coordinate system, and space-time coordinates $x'$, $y'$, $z'$, $t'$ in another coordinate system that is moving to the right with velocity v relative to the first coordinate system, the coordinates are related by the Galilean transformation:

$$\begin{aligned} x' &= x - vt \\ y' &= y \\ z' &= z \\ t' &= t \end{aligned} \tag{2.1}$$

That is, the time is equal in the two systems. This finding contrasts sharply with the important results of the Lorentz transformation and the special theory of relativity, where time in two inertial frames is different (see section 3.3).

Galileo was also the first to attempt to measure the speed of light. Employing a primitive method, he and his servant tried to measure the time it took a beam of light to travel the distance between two hills.[13] Each man stood at the top of one of the hills, holding a lamp that could be covered and uncovered. The servant was told to uncover his lamp the moment he observed Galileo uncovering his. By measuring the time delay, Galileo could then calculate the speed of light. In the acutal experiment the delay decreased to a negligible time as the servant became more efficient. Knowing that the speed of light is 300 million meters per second, we see the inadequacy of Galileo's experimental method.

At the age of 74, under house arrest, Galileo published his masterpiece on mechanics, *Two New Sciences.* Because his publications were eventually outlawed in Italy, the book was smuggled out of the country to be published. In this work he presented the basic principles of mechanics, which were later embodied in Newton's first two laws of motion and the gravitational theory of the universe.

Galileo's celestial discoveries were followed by the work of another great astronomer, Johannes Kepler[5] (1571-1630), a German. Kepler, who devised a model of the rotational movement of planets around the sun, was the first to maintain that their orbits were elliptical rather than circular.

*Johannes Kepler (1571-1630)*

In the early seventeenth century Kepler challenged the visual ray theory. Like da Vinci, he recognized the eye as an elaborate camera obscura, with light entering through the pinhole of the pupil. He was the first to notice that the eye lens focuses an image on the retina. He also proposed that a blurred image results from improper focusing. Kepler further analyzed the operation of thin-lens optical systems, including the Keplerian telescope, which uses a positive eyepiece, and the Galilean telescope, which uses a negative eyepiece. In his book *Dioptrice,* Kepler discussed the total internal reflection and small angle approximation to the refraction laws. In 1625 Christoph Scheiner (1579-1650), a German Jesuit astronomer, removed the eye of an ox and cut away the coating on the back. On the retina he saw an inverted image, thereby confirming Kepler's view of how human vision functions.

In accord with these experimental investigations, the mathematical formulation of optical laws initiated by Kepler began to take a more serious form. In 1621 Willebrord Snell[14] (1591-1626), a Dutch mathematician, formulated the law of refraction, a concept originated by Ptolemy in the second century A.D. Snell derived the relationship between the incident angle (the angle the incident ray makes with a normal to the interface) and the refracted angle (the angle the transmitted ray makes with a normal to the interface). He related the two angles by a quantity he called "the index of refraction of the medium." The index of refraction of a material is the ratio between the speed of light in a vacuum and its speed through that material. This relationship facilitated the mathematical analysis of the interaction of optical rays with lenses and thus spurred the study of optics.

Snell upgraded Ptolemy's observations[3] by naming the observed constant value of the ratio given by the sines of the incident and refracted rays. He used the term "ratios of index of refractions" for the given materials. Snell's equation relating the incident and refracted rays is

$$n_1 \sin\theta_1 = n_2 \sin\theta_2 \qquad (2.2)$$

where $n_1$ is the index of refraction in the first medium,
$n_2$ is the index of refraction in the second medium,
$\theta_1$ is the angle of the incident ray with the normal,
and $\theta_2$ is the angle of the refracted ray with the normal.

In 1637 René Descartes[15] (1596-1650) reasoned from metaphysical precepts that light is a "pressure" (wave) transmitted through the omnipresent "ether." He assumed that ether is a perfectly elastic medium permeating the entire universe through which no energy is lost in transmission. In *La Dioptrique* he argued that the universe is a plenum.* Following Snell, Descartes formulated the modern version of the theory of refraction. He suggested that light travels faster in dense media than through air (a hypothesis later proven incorrect). His work, together with Snell's, formed the foundation of geometrical optics and the basis for analysis of the astronomical telescope. For the first time the location of the image of an object could be predicted by ray-tracing through air and glasses.

*René Descartes (1596-1650)*

Descartes is credited with inventing analytical geometry, and he made many contributions to algebra. His undisputed brilliance, unfortunately, was accompanied by a surprising arrogance. He admitted, for example, to feeling that no secrets in nature were beyond his ability to reveal.

Pierre de Fermat[16] (1601-1665), a French mathematician, was the first to suggest that light slows down when passing through dense media. In 1657 he derived a law of refraction that differed from Snell's. In formulating the law de Fermat assumed a variable "resistance to light" in different media, presenting his "principle of least time," which was similar to the idea originally put forward by Hero in 150 A.D. De Fermat suggested that among all the paths between two points, light travels the path that takes the least amount of time. He

*A plenum is a space entirely filled with matter, a space in which there are no empty subregions.

verified this principle by using rays reflected or refracted from plane surfaces. Later, this principle was shown not to be true for rays reflected from curved surfaces.

The study of phenomena produced by the interference of light rays began with the Italian Jesuit Francesco Grimaldi[17] (1618-1663) in 1660. Grimaldi made the first significant study of the diffraction* of light, a phenomenon whereby light rays apparently bend around obstacles in their path. He observed that the edge of the shadow thrown by a sharply defined object is not correspondingly distinct. Rather, it is surrounded by alternating light and dark bands, decreasing in intensity toward the shadow. This discovery, indicating that the propagation of light is not exactly rectilinear, suggested the existence of a wave motion for light.

In 1665, two British physicists, Robert Boyle[18] (1627-1697) and Robert Hooke[19] (1635-1703), observed a similar effect in the structural arrangement of color rings in thin films, such as those formed by soap or oil floating on water. Hooke suggested that the reflected rays from the front and back surfaces of the thin film cause the colored interference pattern. (The famous "Newton's Rings,"[20] which Newton observed by looking through a convex lens laid upon a flat piece of glass, are also caused by the diffraction effect.) The periodic nature of these phenomena led Hooke to hypothesize that light vibrates through a homogeneous medium such as ether, rapidly generating a new sphere of light at each vibration. This hypothesis was the beginning of the new wave theory of light.

## 2.2.2 Viewing the Invisible: The Invention of the Microscope

Although the amplifying glass was used before the seventeenth century for magnifying small objects, it was not until the middle of that century that a more useful device, the microscope, was invented. The microscope is an ingenious optical device for producing enlarged images of objects that are invisible to the naked eye. The

*Diffraction comes from the word "diffracto," which means "breaking in pieces."

simple one-lens microscope dates from the middle of the fifteenth century when a Dutch student of natural history, Antonie van Leeuwenhoek (1632-1723), developed a low-powered glass magnifier for viewing insects. In *Micrographia,* an essay on microscopes, the English physicist Robert Hooke recorded further early observations of microscopic reality, including studies and illustrations of the crystal structure of snowflakes. His work in microscopic fossils led him to become one of the first proponents of a theory of evolution.

It was not until the 1800s that microscopes were completely free of chromatic aberration.* In 1830 the English surgeon Joseph Lister (1786-1869) discovered that when two separate achromatic lenses are combined to act as a single lens, the image rendered is free of any distortion caused by the curvature of the lens.

A German physicist, Ernst Abbe (1840-1905), was the foremost figure in the development of the microscope during the nineteenth century. He introduced a whole new series of excellent, highly corrected lenses which offered the finest-quality images yet obtainable. His contemporary, a French chemist and microbiologist, Louis Pasteur (1822-1895), was renowned for applying microscopes in his studies of disease-producing organisms, remedies for disease, and the bacteria involved in fermentation.

During the twentieth century, advances in atomic theory and solid-state physics led to the construction of more powerful microscopes. These improvements are discussed in section 4.11, after the discussion of atomic theory.

## 2.3 The Pillars of Science: Formulation of Classical Physics

The second half of the seventeenth century is marked by the accomplishments of two of the most impressive figures in the field of optics, Isaac Newton[21] of England (1642-1727) and Christian Huygens[22] of Holland (1629-1695). During this period Newton pro-

*A chromatic aberration is an inherent defect in a lens, meaning failure to focus all colors at one point.

pounded his revolutionary corpuscular theory of light and color, and Huygens solidified the theory of light propagation on the basis of the wave theory.

In 1665 Newton's studies at Cambridge were interrupted by an outbreak of the plague which closed the university for two years. While residing at home, he purchased a prism at a nearby country fair. By shining light through the crystal he observed that the refracted light reappeared on the opposite wall transformed into a series of rainbow colors:

> *In the beginning of the year 1666 (at which time I applied myself to the grinding of Optick glasses of other figures than Spherical) I procured me a triangular glass-Prisme, to try therewith the celebrated Phaenomena of colours. And in order thereto having darkened my chamber, and made a small hole in my window shuts, to let in a convenient quantity of the Sun light, I placed my Prisme at its entrance, that it might be thereby refracted to the opposite wall. It was at first a very pleasing divertisement, to view the vivid and intense colours produced thereby; but after a while applying myself to consider them more circumspectly, I became surprised to see them in an oblong form; which according to the received laws of Refraction I expected should have been circular.**

Here Newton was demonstrating the fundamental nature of white light, showing that a diffracted light beam disperses into components of well-defined bands of color. By showing that each color had a unique index of refraction, he proved that the effect is a

*Newton, as quoted in Daniel J. Boorstin's *The Discoverers,* Random House, 1983, p. 404 (see reference 5).

fundamental property of light and not of crystal. He also showed that refracting light through more than one prism will not produce colors different from those in the rainbow. On the basis of this experiment Newton developed his theory of optics, which he published in 1672.

*Isaac Newton (1642-1727)*

Newton's theory is based on the idea that light is a concrete substance composed of particles, with different sizes correlated with different colors. Together, the particles blend to form light. They move through space in straight lines and excite the ether into the corresponding vibrational frequency. Red corresponds to the lowest frequency and violet to the highest. Although the results of his experiment could be explained on the basis of the wave theory, Newton chose instead to base them on the corpuscular theory. The main reason for his decision was that the phenomenon of the rectilinear propagation of light could not be explained by the wave theory.

Another of Newton's contributions in optics was his invention of the reflective telescope. The telescopes built by Galileo and others were refractors, which contained large lenses and had chromatic aberrations. Erroneously believing that chromatic aberration was incorrigible, in 1671 Newton focused his attention on using mir-

*Christian Huygens (1629-1695)*

rors in order to avoid this defect. He used two concave mirrors and a flat mirror in his telescopes. The mirrors were smaller than those of the Galilean refractor pieces. Newton's first reflecting telescope was 6 inches long and had a magnification of 40. Later, it was shown how chromatic aberrations in a refractive telescope (a Galilean type) could be corrected.

The interference phenomenon known as "Newton's Rings" was observed when Newton placed a very gradual convex lens with a large radius of curvature on a flat piece of glass. Bands of concentric rings appeared emanating from the central point of contact. Newton catalogued each ring and measured the distance between the two pieces of glass at each ring. He obtained an empirical relationship between these distances and the radii of the rings. This produced a set of concentric rings on the lens emanating from the point at which the plane of the flat piece of glass was tangent to the lens. He referred to this empirical relationship as a "periodic property" of light, but his belief in the emitting particle theory prevented him from recognizing this as a clear example of wave interference. Newton interpreted his finding in a complicated theory of "fits of easy reflection and transmission." Ironically, the length of a fit parameter is the measurement of the wavelength of light observed or the "color" of the ring.

Newton's early studies culminated in 1704 in his comprehensive mathematical treatment of the major properties of refraction, diffraction, interference, polarization, and dispersion (variation of index of refraction with wavelength) in his book *Opticks*. In this plenary work Newton did much to establish and advance the field of optics. Unfortunately, Newton's book inhibited further progress because of his bias toward the discrete particle or "corpuscular" theory of light. Newton's reputation, gained so deservedly through his triumphs in formulating calculus and mechanics, prevented better theories of optics from being taken seriously. His elegantly constructed but fundamentally inaccurate picture of the nature of light dominated serious study for an entire century, until the double pinhole experiment of Thomas Young[23] in 1801.

Newton is also credited with the development of spectroscopy, the science of the analysis of the spectrum of glowing gas. He observed that a gas glows when it is heated or when it is exposed to an electrical discharge. The radiation spectrum of the gas contains intense lines which are unmistakable characteristics of that gas. It was the refinement in spectroscopy and the analysis of radiation line spectra which led to the development of the atomic theory in the first half of the twentieth century.

### 2.3.1 The First Pillar of Physics: Science of Mechanics

One of Newton's major contributions was the derivation of the laws of mechanics, which today are recognized as the laws of classical mechanics. Although the laws of motion were discovered experimentally by Galileo before Newton was born, it was Newton who first formulated them.

First Law: The Law of Inertia

Second Law: Law of Force

Third Law: Law of Action and Reaction

The first law states that, in the absence of any force, if a body is at rest it will remain at rest, or if it is moving at a constant velocity in a straight line it will keep moving at a constant velocity in a straight line.

The second law states that the acceleration of a body, a (the rate of change of the velocity), is directly proportional to the force, F, and is inversely proportional to the mass of the body, m. That is,

$$F = ma \tag{2.3}$$

The third law states that forces produced as a result of interactions between two bodies always appear in equal and opposite pairs. That is, for every action there is an equal and opposite reaction.

These laws undergo some correction when the particle velocity approaches the velocity of light. Nevertheless, Newton's laws ruled the world of mechanics for more than two centuries, until Einstein corrected them for velocities approaching the speed of light.

Newton is considered one of the scientific geniuses of all time and is perhaps the greatest scientist since Aristotle. Indeed, he ranks with Einstein. Although most of Newton's work was published in 1704 in his *Opticks,* he had made most of his major discoveries by the age of thirty. He was famous for the extraordinary discovery of infinitesimal calculus, but his masterpiece is *Philosophiae Naturalis Principia Mathematica* (Mathematical Principles of Natural Philosophy), a revised edition of his earlier work on planetary dynamics. This illustrious work, including his laws of universal gravity, is considered a landmark in modern science.

But genius is often accompanied by pyschosis, and Newton displayed psychotic tendencies throughout his life. A mild example of such tendencies was his inability to confront criticism rationally. He reacted to objections, instead, by flying into rages or withdrawing, for long periods, into virtual isolation.

The major example of Newton's psychosis was his conflict with Gottfried Leibniz. Apart from their disagreements about physics, both claimed credit for discovering calculus. This competition developed into a vengeful war conducted with angry accusations of plagiarism and dishonesty.

Surprisingly, Newton's psychiatric problems did not prevent him from functioning in positions of influence in British society. During his life he served as professor in Trinity College, Cambridge, warden of the mint, and president of the Royal Society.

In 1675 Olaf Romer[24] (1644-1710), through observation of the eclipses of Jupiter's moons, determined that the speed of light is finite, not infinite. He estimated it as $3 \times 10^{10}$ centimeters per second. In experiments with a special crystal he produced a description of double refraction (birefringence)* of calc-spar, observed in 1669 by Erasmus Bartholinus[25] (1625-1698). This crystal has the property of forming two separate images when one looks through it.

In 1678 Huygens strengthened the wave theory of light mathematically. He proposed that, if a point of the ether is illuminated by a ray of light, it produces a disturbance, or "wavelet," that fans out in all directions, forming a spherical wave. The "envelope" created by the common convolute of all these wavelets becomes a new "wave front" of the light ray. From the geometric construction of this principle, Huygens derived the laws of reflection and refraction. He concluded that the index of refraction of any medium is the ratio of the speed of light in a vacuum to its speed in that medium. In 1690 Huygens described ether as an all-pervading medium composed of tightly packed elastic balls. He argued that light is carried by this medium in waves.

## 2.3.2 A Fight over Light: Particle Theory versus Wave Theory

Newton's particle theory and Huygens's wave theory com-

*"Birefringence" means refraction of light into two slightly different directions.

peted furiously for scientific acceptance until the mid-nineteenth century, when James Maxwell combined light with electromagnetic waves and generated his celebrated electromagnetic equations. Both Newton's and Huygens's works, however, helped to remove the obsolete visual ray theory from the realm of scientific debate.

Huygens is also noted for his discovery of the phenomenon of polarization.* He passed two image rays obtained from one calc-spar crystal through another crystal rotated at a right angle to the first. This process extinguished both images, owing to the fact that no light passes through two polarizers crossed at right angles.

*Thomas Young (1773-1829)*

By this time Huygens had developed the most specific expression of a light-wave theory. Ironically, his demonstration of the polarizability of light became the basis for Newton's rejection of such a light-wave description. Newton's conclusion sprang from the fact that physical concepts of wave propagation were mainly supported by the physical properties of sound. In sound, the transfer of disturbance takes place along the direction of propagation, forming longitudinal waves. The "transversality" implied by the calc-spar extinction of

*When the electric field of light is oriented in a specified way with respect to the propagation direction, light is called polarized. When the electric field is oriented in an unpredictable way, light is called unpolarized. Light from a rainbow is, for example, polarized. Most light sources are partially polarized. The human eye is unable to detect the presence of polarization.

images could not be explained by using a longitudinal wave model. This quandary caused Newton to reject light-wave theory and to develop his atomistic "corpuscular" theory. His conception was that particles of light are continually emitted radially from tiny luminous corpuscles speeding along the direction of propagation.

*Augustin-Jean Fresnel (1788-1827)*

In 1801 the British physicist and physician Thomas Young (1773-1829), by studying the power of accommodation, discovered how the eye's lens changes shape during focusing. Turning his attention to light, he devised an ingenious experiment which reinforced Huygens's wave theory. Young passed light through two pinholes in a screen, letting it fall on a second screen placed behind the first. Upon illuminating a screen with two pinholes of light, he obtained diffraction patterns that consisted of alternating light and dark bands. He showed that in the region of overlap between the two beams, constructive (light) and destructive (dark) interference was taking place. This phenomenon could be explained only on the basis of the wave theory. Newton's discrete particle theory would not have exhibited the same patterns. A musician as well as a scientist, Young knew that sound waves of different frequencies could strengthen or could cancel each other out. His only explanation for the dark lines formed on the screen was that light waves from the different holes cancel each other out.

When he measured the waves Young found that different colors have different frequencies, red having the lowest and violet the highest. Attempting to account for color vision, Young theorized that the retina harbors sensors that respond to different frequencies and hence to different colors of light. By experimentation he showed that all colors could be reproduced by combining red, green, and violet. Young therefore concluded that the retina needs only three types of sensors corresponding to these colors to produce color vision. In 1817 he proposed that the waves from which light is formed are transverse to the direction of propagation of the beam. This idea embodied the first satisfactory explanation of the polarization of light (since waves vibrating at right angles to their direction of travel could continue to vibrate even when restricted to one plane parellel to that direction).

*Dominique-Francois Arago*
*(1786-1853)*

So pervasive was Newton's influence, however, that Young's efforts attracted only ridicule, not commendation. When harmful publicity began to affect his medical practice, he abandoned his investigations. The results of Young's scientific work, however, were reconfirmed when more extensive diffraction experiments and concise mathematical analyses were performed by the French physicist, Augustin-Jean Fresnel[26] (1788-1827).

Young's crucial contribution to the wave theory of light was his assumption that light is a transverse* rather than a longitudinal† wave. He based his idea on experiments with polarized light conducted by the French scientist, Dominique-Francois Arago (1786-1853), and Fresnel[27] in 1816. Fresnel and Arago found that two polarized rays at right angles to each other never interfere with each other, a discovery in conflict with the assumption that light is a longitudinal wave.

*Pierre-Simon de Laplace (1749-1827)*

The classical particle controversy continued until the proponents of the corpuscular theory, encouraged by advances made by Pierre-Simon de Laplace (1749-1827) and Jean-Baptiste Biot[28] (1774-1862), suggested that the 1918 prize question sponsored by the Paris Academy be to find a solution to the problem of diffraction. The implication of their proposal was that the successful solution would be based upon and would confirm the "emitting corpuscle" theory. To their great surprise and dismay, the award went to

*Transversality of the field means that the electric and magnetic field vectors lie in planes normal to the direction of propagation such as in light.

†Longitudinal disturbance means that the disturbance in the field is in the same direction as the direction of propagation such as in sound.

Fresnel, whose treatment of diffraction was based on Huygens's wavelet envelope method and Young's interference principle. Fresnel's successful analysis of the diffraction phenomenon, based on his refinements of Huygens's idea of secondary spherical wavelets, included proposing an "obliquity factor" which concentrated the intensity of the wavelet toward the direction of propagation of the wave front. This factor was essential in correcting Huygens's model because of the fact that no backward-directed wave was observed. Fresnel's analytic technique was to split the individual spherical wave fronts into concentric rings of "half-period zones." The new wave front at a point in the forward direction of propagation was obtained by in-phase contributions of the concentric rings of the spherical wave.

Fresnel's brilliant application of these concepts led to concise descriptions of a number of phenomena. In 1818, on the basis of Fresnel's work, Siméon-Denis Poisson[29] (1781-1840), a French mathematician, put forward an argument in favor of Newton's corpuscular theory. He argued that if a shadow of a perfectly round object is made by a point source, then at the periphery of the shadow all the waves must be in phase and a bright point must be present. When Fresnel and Arago carried out the experiment, a bright point did appear at the center. It was another triumph for the wave theory.

In 1818, inspired by Young's idea of transverse vibrations of optical waves, Fresnel evaluated the influence of the earth's motion on the propagation of light. The question was whether or not light from stellar sources differed from light from terrestrial sources. On the basis of experimental investigations by some of his contemporaries, Fresnel developed his theory that the earth causes partial ether drag (see section 2.4).

Fresnel also initiated the analysis of dispersion phenomena* in terms of a prism's molecular structural interaction with light

*The separation of an electromagnetic wave into its various frequency components is called dispersion. A beam of white light, for example, can be separated into its monochromatic components because of the different velocities of rays at different wavelengths of the beam as it passes through a prism. The dispersion of a material, such as glass or water, at a given wavelength is defined as the rate of change of refractive index with wavelength.

waves. Fresnel's work was further elaborated by the French mathematician Augustin-Louis Cauchy[30] (1789-1857).

The most convincing evidence against the corpuscular theory of light was presented in 1850. According to the corpuscular theory, the corpuscles' attraction for dense media causes refraction (implying a faster speed of light in denser media). Jean Foucault[31](1819-1868), Armand Fizeau (1819-1896), and Brequet[32] observed that light actually slows down while passing through water (as compared with air), a phenomenon that is consistent only with the wave refraction theory.

*Joseph von Fraunhofer (1787-1826)*

In measuring the speed of light, Fizeau and Foucault used a method similar to Galileo's except that they replaced the servant with mirrors. Instead of the on-and-off cover of the lantern, Fizeau used a rotating chopper wheel and Foucault relied on a rotating mirror. Foucault also measured the velocity of light in water as well as in air. He found a lower velocity in water, thus supporting the wave theory of light.

Joseph von Fraunhofer[33] (1787-1826), a contemporary of Fresnel's, made his contribution to the wave theory by examining a special case of diffraction patterns, when an object is situated at a great distance from the optical instrument.* Fraunhofer's pattern has been a convenient means for the modern development of the optical spatial filtering technique.[34]

### 2.3.3 Lights and Shadows of the Elements: The Secrets of the Atom

Fraunhofer is also noted for establishing the phenomenon of absorption lines, known as Fraunhofer lines. After Newton initiated the study of spectroscopy, William Wallastron (1766-1828) continued it, discovering dark lines in the solar spectrum in 1802. From 1814 to 1817 Fraunhofer pursued the study of these absorption lines in the solar spectrum, finally obtaining the first wavelength determination of the Fraunhofer lines. In 1861, more than forty years later, Robert Bunsen (1811-1899) and Gustav Kirchhoff[35] (1824-1887), experimenting with the solar spectrum, reconfirmed the existence of Fraunhofer lines. Their experiments indicated that the continuous spectrum emitted from the sun's core loses radiation in some of its wavelengths

*Gustav Kirchhoff (1824-1887)*

*Fraunhofer showed that, for a monochromatic light, the complex amplitude of the diffraction pattern is related to the Fourier transform of the entrance aperture. Even in the presence of aberration, a modified Fourier transform of the entrance aperture is related to the complex amplitude of the diffraction pattern.

as it passes through the sun's atmosphere. Bunsen and Kirchhoff thus concluded that each atom has its own characteristic wavelength in emission and in absorption.

These observations, which elucidated the process of line absorption and emission, appear to be of much greater significance in the development of atomic theory than do the observations made from Maxwell's equations (see section 3.4). The intriguing question of how light and matter interact led to the theoretical development of the quantum theory in the twentieth century. The experiments of Fraunhofer, Bunsen, and Kirchhoff also gave impetus to further work in spectroscopy, through which the line spectrum of any chemical element can be detected and analyzed.

A Swiss mathematician, Johann Balmer[36] (1825-1898), continued the study of spectroscopy. In 1885 he discovered that the spectral lines of hydrogen are discrete and are denser for shorter wavelengths. He also expressed the optical frequency of the absorption lines, $\nu$ , in a simple equation:

$$\nu = R\left(\frac{1}{4} - \frac{1}{n^2}\right) \qquad (2.4)$$

R, a constant, is named the Rydberg Constant after Johannes Rydberg (1881-1919), a Swedish physicist. The value of R is 109,677.58 $cm^{-1}$. When n = 3, the Balmer series produces the red hydrogen line corresponding to $\lambda$ = 0.656 micrometer.

Theodore Lyman (1874-1954), an American physicist, continued the investigation, discovering, in 1906, spectral lines of hydrogen similar to those found by Balmer, but in the ultraviolet region. Soon afterward, Friedrich Paschen and A. H. Pfund of Germany and F. S. Brackett of the United States found more spectral lines in the infrared region. The Balmer formula was modified to make it fit all these lines, and a general formula was developed for line absorption of hydrogen:

$$\nu = R\left(\frac{1}{m^2} - \frac{1}{n^2}\right) \qquad (2.5)$$

where m is allowed to have values, 1, 2, 3, . . . and n is allowed to have values (m + 1), (m + 2).... Although the physical origin of these lines and their corresponding equations remained unknown until development of the atomic theory in the early twentieth century, knowledge of the spectral locations of the lines in the hydrogen atom was crucial for testing the hypothesis of an atomic model at the time.

Herman von Helmholtz (1821-1894), a German scientist, confirmed Young's theory of transverse vibrations of optical waves in 1850. He also made a valuable contribution to the study of vision. Because one cannot peer into an eye without blocking the light, Helmholtz built a device called the ophthalmoscope, which enabled him to observe the eye without obscuring the view. For the first time, looking through this instrument, he saw the tiny blood vessels on the surface of the retina, vessels that nourish the cells that make vision possible.

Humphrey Lloyd[37] (1800-1881), a British scientist, verified Fresnel's polarization theory in 1853 by observing conical refraction. Following up on predictions by William Rowan Hamilton[38] (1805-1865), an Irish mathematician, Lloyd confirmed Fresnel's observations by using his transverse wave formulation.

Before discussing the discovery of the fundamental relationship of electricity, magnetism, light, and Maxwell's equations, it is appropriate to describe the advances made during the golden century of electricity and magnetism, 1775-1865.

## 2.3.4 The Golden Age of Electricity and Magnetism

During the late 1700s, as experimental investigations in optics were beginning to bear fruit, major advances were being made in the field of electricity and magnetism. Fundamental discoveries and innovations in all three fields in the period 1775-1865 culminated in the reformulation of the classical equations of electromagnetism.

In 1785, after a series of impressive experiments, Charles-Augustin de Coulomb (1736-1806), a French physicist, formulated a law stating that the force between two electric charges is proportional to their product and inversely proportional to the square of the distance between them. He also showed that the force exerted on one charge is the summation* of forces exerted by other charges present. Henry Cavendish[39] (1731-1810), an English physicist, introduced the idea of "electrical potential"† in formulating the laws of electricity. Struck by this idea, Siméon Poisson began experimenting with the concept of an electric field;‡ in 1813 he identified the spatial derivation of the electrostatic potential as the electric field vector.

*Charles-Augustin de Coulomb (1736-1806)*

The next advance was made by the Danish physicist and chemist Hans Christian Oerstèd (1777-1851), who discovered in

*Summation here refers to vector summation, in contrast with scalar summation.

†Electric potential is defined as the relative voltage at a point in an electric circuit or field with respect to some reference point in the same circuit or field.

‡Electric field is a condition in space surrounding an electrically charged body such that the forces due to the charge are detectable.

*Henry Cavendish (1731-1810)*

1820 that an electric current flowing through a wire could deflect a magnetic needle. His experiment was crucial to electromagnetism because it showed that a flow of electricity produces a magnetic effect. Within a week of Oerstèd's discovery, André-Marie Ampère (1775-1836), a French mathematician, developed a theory for this new phenomenon. Known as Ampère's Law, it expressed mathematically the magnetic force between two electric currents. Ampère demonstrated that two adjacent wires, given freedom of movement, attract each other if their electric currents flow in the same direction. Conversely, the same wires repel each other if the currents flow in opposite directions.

Soon thereafter, in 1827, the German physicist Georg Ohm[40] (1787-1854) found that a current flowing through a conductor is proportional to the electrical potentials at the two ends of the conductor and inversely proportional to the electrical resistance between the two ends of the wire. Although his discovery was given a cold reception by the scientific community of the time, it was eventually acknowledged as true, and is now known as Ohm's Law. Later, Karl Friedrich Gauss[41] (1777-1855), a German mathematician, physicist, and astronomer, advanced the potential theory by formulating mathematical equations relating the electric field on a closed surface to the total electrical charge enclosed by this surface.

Michael Faraday [42] (1791-1867), an English pioneer of science and one of the most imaginative experimentalists of all time, is known for two simple but enlightening discoveries. Aware of Oerstèd's experiment and convinced of the interrelationship between electricity and magnetism, he demonstrated in 1831 the phenomenon of electromagnetic induction and the generation of an electric current by a change (time variation) in the magnetic field. Faraday's experimental device consisted of two magnetic bars and a wire through which an electric current flowed. The current passing through the wire caused the magnetic bars to rotate. An alternative method led to the same result. When the magnetic bars are fixed, the current passing through the wire causes the wire to rotate around the magnet. Faraday hypothesized that the current passing through a wire introduces a flux line or a magnetic field. He was thus the first to show that electricity and magnetism, although of different manifestation, originate from the same source. Faraday's experiments in electricity were the first to be applied to the practical mechanical and/or electrical innovations that

*André-Marie Ampère*
*(1775-1836)*

would later revolutionize human life, redefine the priorities of science and industry, and endow life with new meaning.

### 2.3.5 The Second Pillar of Physics: Unified Electromagnetic Theory

Although the fields of electricity, magnetism, and optics seemed disjointed around 1760, a century later it had become apparent that a vital connection existed among them. For more than the preceding two centuries, research in optics had persisted, despite a major controversy over the wave and the corpuscular theories of light. The remarkable advances that were simultaneously being made in the fields of electricity and magnetism eventually led to their merger into the field of electromagnetism. The mathematical formulas for electricity and magnetism which had been developed during this period were proven by experimentation to hold true. The experiments suggested that electricity and magnetism, because of their common properties, might originate in the same source. The stage was set for the three major fields to meet and combine forces, leading to the formulation of a unified and all-encompassing theory.

The Scottish physicist James Clerk Maxwell[43] (1831-1879), recognized as a genius in his own time, was the one who succeeded in unifying these seemingly different disciplines. His brilliant theory gave mathem-

*Georg Ohm (1787-1854)*

atical form to several of the most fundamental physical and empirical relationships between electricity and magnetism. He expressed this theory in a single set of equations.

*Karl Friedrich Gauss (1777-1855)*

Inspired by Faraday's line of thinking, Maxwell set himself to the task of discovering whether a basic relationship existed between the electric and the magnetic fields. He based his work on Coulomb's Law, Gauss's Theorems, Faraday's Law, and Ampère's Law. While formulating and systematizing his equations, Maxwell realized that a significant term was missing from Ampère's Law. Applying a symmetrical approach, he hypothesized that the generation of a magnetic field when an electric field has a time variation is analogous to the generation of an electric field when a time-varying magnetic field exists. Discovering the term missing from Ampère's Law, he called it the displacement current. By integrating this term into his formulas, Maxwell produced his remarkable electromagnetic equations.

Maxwell unified all previous experimental results and their corresponding equations into a set of equations which took the form of a wave equation. He deduced the speed of propagation of this wave, c, from purely electrical and magnetic constants.* It is of particular importance to note here that in 1856 Wilhelm Eduard Weber (1804-1891) a German physicist, and R. Kohlrausch[44] (1809-1858) had discerned the relationship between the electric permittivity and magnetic permeability constants, with c the speed of light, portending the association between electromagnetic radiation and light. Instead of accepting this relationship as a coincidence, Maxwell inferred that light is also an electromagnetic wave. This was a crucial turning point in the history of optics.

*Michael Faraday (1791-1867)*

## 2.3.6 Triumph of the Wave Theory: Maxwell's Equations

Maxwell's remarkable set of four equations describe all the phenomena of electricity and magnetism. They demonstrate that electricity and magnetism are not separable. *Whenever a time-varying electric field exists, there must be a magnetic field, and whenever a time-varying magnetic field exists, there must be an electric field.* Maxwell's equations,[45] relating the electric field intensity E, the magnetic field intensity H, the electric displacement D (flux density), the magnetic flux density B, the volume charge density $\rho$, and the current density J, are expressed in MKS units as:

* $c = 1/\sqrt{\epsilon_0/\mu_0}$ where $\epsilon_0$ and $\mu_0$ are the permittivity and permeability of free space, respectively.

$$\nabla \times E = -\frac{\partial B}{\partial t} \qquad \text{Faraday's Law}$$

$$\nabla \times H = J + \frac{\partial D}{\partial t} \qquad \text{Ampère's Law} \qquad (2.6)$$

$$\nabla \cdot D = \rho \qquad \text{Gauss's Law}$$

$$\nabla \cdot B = 0$$

These equations also indicate that a change in the electric field induces a change in the magnetic field, which in turn induces a change in the electric field, and so on. The two fields generate each other, and the resulting electromagnetic field progresses outward in all directions. On purely theoretical grounds, Maxwell demonstrated that an electromagnetic wave propagates as a transverse wave.

Maxwell's equations produced dramatic results. Light was proven to represent a disturbance that occurred in the form of an electromagnetic wave. These equations, rich in content, which rank in importance with Newton's gravitational equation, are of a different form. They are unique, for they describe a new type of physical phenomena. Their

*James Clerk Maxwell (1831-1879)*

distinctive features appear in all equations of modern physics. Einstein called the formulation of Maxwell's equations "the most important event in physics since Newton's time." An admirer of Faraday and Maxwell, Einstein also noted that "just as Newton has given mathematical form to Galileo's insight, Maxwell performed similar functions for Faraday."

In spite of his rigorous approach, Maxwell's theory was not immediately accepted. He and his followers persistently tried to describe the electromagnetic field with the aid of mechanical models. The search for an "explanation" of Maxwell's equations in terms of such models was eventually satisfied by the extensive experiments conducted by Heinrich Hertz[46] (1857-1894), a German physicist. In 1888 Hertz generated and detected radio waves,thus finally verifying the existence of electromagnetic waves. From that point on, the wave theory of electromagnetism and Maxwell's equations were widely accepted. Today they are one of the main pillars of classical physics.

*Heinrich Hertz (1857-1894)*

Maxwell's equations forced the needed conclusion that light has an impact on matter. Because of the electric and magnetic fields of light, it was suspected therefore that atoms must also have electric and magnetic properties. The investigation of atoms followed, leading subsequently to an understanding of the microscopic nature of light. At that point the study of quantum optics commenced.

In 1876 Gustav Kirchhoff, using a superior method of integration than had previously been employed, corrected two deficiencies in Fresnel's wave theory, the obliquity factor and the phase of the spherical wavelets. Later, Albert Abraham Michelson (1852-1931), a German-born American physicist, devising a delicate experimental apparatus, refined the calculation of the speed of light by about 0.03 percent.[47]

## 2.4 Cracks in the Pillars: Failure of Classical Electromagnetism

After Maxwell had derived his electromagnetic equations it was assumed that optical phenomena could be completely analyzed and described by application of the equations. At the end of the nineteenth century, however, it was realized that the electromagnetic theory failed to explain the photoelectric effect.

In 1887 Hertz[48] showed that if a beam of light is directed onto a clean metal plate in a vacuum, electrons are emitted from the plate. The kinetic energy of the electrons escaping the metal depends on the frequency of the incident beam and is independent of the beam's intensity. Hertz's observation conflicted with the classical theory of electromagnetism, which holds that, if radiation is absorbed continuously, electrons should be able to absorb more energy from more intense beams. The conflict between Hertz's experimental results and the theory of electromagnetism lasted until 1900. At that point the experimental results of another phenomenon, blackbody radiation, were also shown to conflict with the prevailing electromagnetic theory.

### 2.4.1 The Ether Controversy

Hertz's experiments firmly established the fields of optics and electromagnetism on the basis of the wave theory set forth in Maxwell's electromagnetic equations. The acceptance of this theory by the mechanically minded physicists of that time depended on their acceptance of the idea of an all-pervasive medium within which light waves could propagate. Because any wave motion, such as sound, had previously been experienced in this manner, it was assumed that these light waves also required a medium through which to propagate. But for sound waves, propagation consisted of an amplitude oscillation in the direction of propagation (longitudinal waves), whereas the electric and magnetic fields for electromagnetic waves were perpendicular to the direction of propagation (transverse waves). In view of the known facts about light, it was necessary to assume that the medium, then referred to as ether, permeated all space, was of negligible density, and had negligible interaction with matter. Presumably, it existed solely for transporting electromagnetic waves.

The concept of ether originated in the word "quintessence" (derived from the Latin "quinta essentia")[5] which means fifth element. According to an ancient Greek belief, the four elements (earth, water, air, and fire) occupy part of the universe and ether, the fifth element, fills up the remaining part, where the substantive elements do not exist.

The next step, after Maxwell's electromagnetic theory had been accepted, was to determine the physical nature of ether. Such a medium could be expected to possess strange properties. It should support a light speed of $3 \times 10^8$ meters per second, and it should support the vibrations of light of more than $10^{10}$ hertz. It was also assumed that ether did not interfere with the motion of the heavenly bodies. (By this time it was known that all celestial bodies are in motion.)

### 2.4.2 Doubting the Absolute

According to Newton, the fabric of space, ether, is at absolute rest. The absolute motion of an object, on which Newton's law of motion is based, was then determined from an absolute coordinate at rest with ether. However, revelations emerging slowly over the next two centuries provided evidence for believing that ether does not exist and that there is, therefore, no absolute coordinate system at rest with it. It followed from this that the laws of classical mechanics had to be modified.

In 1725 James Bradley[49] (1693-1762), an English astronomer, tried to measure the distance to a star by a method of triangulation. The orbital motion of the earth and its position in orbit provided a large base for triangulation. He discovered, however, that fixed stars showed an apparent displacement related to the direction of the earth's motion and not to its position in orbit. These apparent changes in direction were shown to be similar to the apparent angle of raindrops relative to a moving observer.* Thus Bradley's experiment indicated that ether remains fixed as the earth moves through it.

*Albert Abraham Michelson (1852-1931)*

Fresnel had assumed that ether, while at rest, permeates the opaque earth and remains unaffected by its motion, but experimental investigations in 1838 by Dominique-Francois Arago proved otherwise. Observing light beams from nearby and distant sources,

*For an observer in motion, the vertical raindrop appears at an inclined angle.

such as starlight, through a glass, Arago found no difference between light beams from celestial objects and those from terrestrial objects. He concluded that the earth is at rest with respect to ether. Subsequently, Fresnel derived a theory based on that conclusion, a theory that was verified experimentally between 1849 and 1851 by Armand Fizeau. Fizeau speculated that ether is partly, and only partly, dragged alongside a moving, transparent substance. In the experiment, he used an optical apparatus in which water flowed through two parallel tubes. To evaluate the effect of water velocity on the speed of light, Fizeau passed a light beam through the water both in the forward direction and in the opposite direction of the water flow. He concluded that light behaves in the same manner if any fraction of velocity of moving water is added to light velocity. Thus he confirmed Fresnel's hypothesis. Both Fizeau's experiment, and George Airy's (1801-1892) experiment in 1871, which examined stellar aberration using a water-filled telescope, confirmed Fresnel's drag hypothesis. But all subsequent investigations, notably those of A. Michelson and the American chemist Edward William Morley[50] (1838-1923) in 1887, failed to measure the required drift.

In 1879 Maxwell suggested a scheme for measuring the speed at which the solar system moves with respect to ether. He did not live long enough, however, to see the results. In 1881 Michelson used Maxwell's scheme to construct the delicate interferometer, an ingenious approach for measuring that speed. Michelson's idea was to measure the time duration of a light beam traveling to and reflected back from a mirror located at a known distance. If the mirror is so located with respect to the light source that the beam is directed in the same direction as the earth's motion, it should take longer than if the beam is directed in a direction perpendicular to the earth's motion. It was easily shown that if the earth's velocity through ether is v, then the time ratio is $(1 - v^2/c^2)^{1/2}$. Although the velocity, v, of the earth relative to the ether would be much less than the velocity of light relative to ether, Michelson found no detectable time change in light traveling in different directions.

In 1887 Michelson and Morley repeated the same experiment with greater precision. They measured the velocity of light moving in various directions compared with the earth's motion. Still no differences were detected. This experiment has been repeated many times since Michelson and Morley first performed it, with various modifications and greater accuracy. As late as 1964 a still more sensitive test was made by T. S. Jaseja,[51] who placed two laser beams at right angles to each other. Once again the result confirmed the Michelson-Morley findings. No measurement has yet shown the motion of the earth relative to ether. The conclusion from Michelson and Morley's experiment was inevitable: either the ether moves with the earth or there is no ether at all. In either case, the concept of absolute motion in classical physics and Newton's concept of the universe, based on the assumption of a medium at absolute rest, were both shaken.

*Hendrik Antoon Lorentz (1853-1928)*

The ether-drag hypothesis of Fresnel, although it could explain the Michelson-Morley experiment and did not require modification of existing theories in classical mechanics, nevertheless could not explain the stellar aberration discovered by Bradley.

In 1895 George Francis Fitzgerald (1851-1901), an Irish physicist, attempted to explain the results of the Michelson-Morley experiment while keeping intact the equations of mechanics and electromagnetism. Proposing that the null effect of the Michelson-Morley experiment was due to a compensating factor, he suggested that the length of every object is reduced in the direction in which the object is moving (known as the foreshortening effect), whereas the length of an object remains constant in the direction perpendicular to the motion. Thus, if the velocity of an object is v, the amount of reduction is a factor of $(1 - v^2/c^2)^{1/2}$. The foreshortening fraction due to the earth's velocity cancels the factor $(1 - v^2/c^2)^{1/2}$, which related the velocity of light in the Michelson-Morley experiment in two perpendicular directions. For speeds much lower than c, the foreshortening factor is negligible. For a body moving at the speed of light, however, its length in the direction of motion becomes zero, leading to the assumption that no velocity higher than the speed of light is possible.

In 1985 Hendrik Antoon Lorentz (1853-1928), a Dutch physicist, introduced the idea of local time (different time zones in different locations). Influenced by Fitzgerald's proposal that moving bodies approaching the velocity of light contract in the direction of motion, Lorentz expanded his idea in 1894 and arrived at the Lorentz transformation. He found that the Galilean transformation altered the format of Maxwell's equations, but he insisted that his own transformation left Maxwell's equations intact.

The Lorentz transformation relating two inertial frames* x and x′, one moving with the velocity v with respect to the other, is

$$x' = \frac{x - vt}{\sqrt{1 - \frac{v^2}{c^2}}} \qquad t' = \frac{t - \frac{vx}{c^2}}{\sqrt{1 - \frac{v^2}{c^2}}} \tag{2.7}$$

*Inertial reference frames refer to those frames in which a body, free of forces, moves with constant velocity.

A major contribution of the Lorentz transformation was the discovery of the relationship between space and time through the velocity of light. For velocity much lower than that of light, the Lorentz transformation reduces to the Galilean transformation. Lorentz proposed a theory dictating that the mass of a particle with a given charge is inversely proportional to its radius. In other words, the smaller the charge of a particle, the greater its mass. Lorentz also showed that the mass of a particle in motion, m′, is related to its rest mass, m, by the equation $m' = m(1 - v^2/c^2)^{-1/2}$, which resembles Fitzgerald's foreshortening fraction. Lorentz's work later formed the basis for Einstein's special theory of relativity.

After the Michelson-Morley experiment, it was harder to maintain the ether theory and the laws of classical mechanics. For centuries the laws of mechanics, as initiated by Galileo and advanced and formulated by Newton, were the main pillars of science. These laws remained invariant for various coordinate systems moving in a uniform motion with respect to one another (the so-called inertial reference frame). By the end of the nineteenth century, however, two main problems had surfaced:

1. Experimental verification of the constancy of the speed of light in various frames of reference was impossible by any ether hypothesis, the ether-drag theory, or even the Fitzgerald Lorentz contraction theory.
2. The form of Maxwell's wave equations was changed by the Galilean transformation and therefore became unacceptable. The fact that the description of physical phenomena must be independent of their frame of reference made it apparent that something was wrong with the Galilean transformation of Maxwell's equation.

In 1899 Jules-Henri Poincaré (1854-1912) expressed his doubt of the existence of ether. The ether hypothesis could not explain the Bradley, Fizeau, Airy, and Michelson-Morley experiments simultaneously. No modification of this hypothesis, such as Fresnel's partial drift of ether or the Fitzgerald-Lorentz contraction, could explain the results of all these experiments.

By the beginning of the twentieth century the field of physics was in turmoil. Uncertainty loomed over the foundations of classical mechanics, classical electromagnetism, and the Galilean transformation. The stage was set for a radical change; a new theory was needed.

### 2.4.3 Ultraviolet Catastrophe

Concurrent with the problems surrounding the existence of ether and Maxwell's equations under the Galilean transformation was the problem of black-body radiation, which was just surfacing at the end of the nineteenth century.

By the mid-nineteenth century it was well known that an object, when heated, radiates energy. Depending on the temperature of the object, radiation wavelengths extend from visible to ultraviolet spectra, yet they always contain the infrared portion of the spectrum. In 1859 Gustav Kirchhoff[52] indicated that the radiation spectrum of an object is independent of the shape, size, and chemical properties of the material of that object. Twenty

*Wilhelm Wien (1864-1928)*

years later Josef Stefan (1853-1893), an Austrian physicist, showed experimentally — and in 1884 Ludwig Edward Boltzmann (1844-1906), another Austrian physicist, showed theoretically — that the radiation energy of a blackbody* increases as the fourth power of its temperature. The German physicist, Wilhelm Wien[53] (1864-1928), discovered in 1893 that the distribution of this energy is a function of the radiation wavelength. He derived the function from the second law of thermodynamics.

Certain perplexing problems of blackbody radiation arose during the latter part of the nineteenth century. The experimental results of blackbody radiation were quite different from the prediction of the reigning theory at that time, particularly in the ultraviolet portion of the spectrum. This deviation was called the "ultraviolet catastrophe." In 1896 Wien himself reasoned that the distribution function for a blackbody is exponential. Although Wien's exponential curve complemented the experimental data, the reasoning behind his proposal needed significant improvement.

## 2.5 Looking through the Human Body: Discovery of X Rays

Another important optical discovery made at the close of the nineteenth century revolutionized the fields of medicine and biology in the twentieth century. In 1895 Wilhelm Conrad Röntgen (1845-1923) a German physicist, discovered what is known today as the X ray. He found that if a surface coated with barium platocyanide is placed outside a cathode-ray discharge tube, it emits a light (fluorescence). He deduced that when the electrons of a cathode-ray tube struck the glass wall of the tube, an unknown radiation was produced. This radiation traveled toward the chemical surface, striking it and causing the fluorescence. Since this strange new ray did not exhibit

*A blackbody is defined as an object that absorbs light but does not reflect or transmit the incident radiation. In the study of radiation, the blackbody must always be considered, as thermal radiation from a black object is the most intense form of radiation.

any common properties of light, such as reflection or refraction, Röntgen mistakenly did not relate these rays to light. Because of its unknown origin, he named it the X ray. Röntgen used his wife's hands as the subject for the first X-ray image, revealing the intricacies of bone structure in the human body. His wife, however, hated the image and called it the "ghost image," as the view of a skeleton was associated with death.

Today, X-ray diagnosis is unparalleled by any other diagnostic tool in medicine. These miracle rays reveal the functioning or malfunctioning of the various parts and organs of the body. Without them, no one can guess where medicine might be today.

When X rays were applied to crystallography, in 1953, the second scientific mystery of our century (after the atom) was unlocked. DNA and RNA molecules were discovered and found to be the building blocks of living organisms.

# 3

# The Explosion of the Atom: The First Half of the Twentieth Century

## 3.1 Introduction

During the three hundred years from the seventeenth century to the beginning of the twentieth century, the three main pillars of physics — mechanics, electromagnetism, and thermodynamics — were firmly established and had been tested experimentally. Both the foundation of mechanics by Newton, following Galileo's experimental investigations, and the formulation of electromagnetism by Maxwell, following Faraday's experiments, were solidly instituted. These key formulations, supported by experiments, were considered to be the foundation of science. As such, they seemed unshakable.

By the end of the nineteenth century, however, the solidity of the laws of mechanics and electromagnetism was being questioned. After the Michelson-Morley experiments, the existence of an ether, an all-pervading medium in which light had been thought to propagate, remained in doubt. And the problem associated with the constancy of physical phenomena in various reference frames was also unresolved. An outstanding example of this last problem was the form of Maxwell's electromagnetic equations under the Galilean transformation, upon which the laws of classical mechanics were based. In addition, other problems were surfacing. Puzzling questions, such as the blackbody radiation spectrum and the interaction between light and matter known as the photoelectric effect, still remained unanswered. The physical process behind the absorption and emission of light by or from matter was unexplained. Conflicting theories of light were subject to constant dispute. The classical theory of mechanics and electromagnetism was uprooted by new discoveries and more innovative theories. Thus the early scientific period bequeathed monumental problems for turn-of-the-century scientists to resolve.

The next era spawned a new breed of physicists, many of them the most ingenious scientists of all time. The first of many scientific milestones in this period was the resolution of the laws governing blackbody radiation. This advance opened entirely new avenues to the solution of other nineteenth-century problems, eventually providing the necessary background material for quantum mechanics, the most significant theory in physics developed in the twentieth century. As we shall see, applied mathematics became the key to uncovering and understanding the profound laws of nature.

This section presents the crucial developments of the first half of the twentieth century. It sets out the mathematical arguments that were most important to those developments, along with a simple explanation of what emerged during the years that followed. The material between pages 81 and 133 contains many equations but is restricted to only those equations absolutely necessary for an understanding of the period. Still, readers who are not well equipped for so technical a discussion may prefer to skim this section. The present introduction should provide an overview sufficient to make later sections understandable.

A fundamental transition occurred in the minds of scientists as physics underwent its metamorphoses. In order for these new laws and theories to be conceptualized, the character of thought and of experimentation had to change. They had to become probabilistic rather than deterministic.

The first breakthrough came in 1900, when Max Planck gave an accurate description of blackbody radiation, based on the theory that the energy levels in an atom are quantized. Planck went on afterward to explain blackbody radiation as the radiation produced by the loss of energy when an electron makes a transition from one energy state to a lower one in a blackbody.

About five years later, Albert Einstein also explained the process of light absorption and emission by matter, by assuming that atoms had discrete rather than continuous energy levels. Planck's and Einstein's concepts of the emission and absorption of optical radiation and of discrete energy levels were the springboard for the development of quantum mechanics.

Concurrent with the discovery of the photoelectric effect, the question of the existence of an ether was also answered. In a paper written in 1905, Einstein made the daring assumption that the ether does not exist. Until that time it had widely been believed that ether filled up any empty space in the universe and was the drifting element that carried light or energy throughout the universe. Einstein's assumption cleared up many contradictions between experimental results and this antiquated belief.

Next, the laws of classical mechanics under high-speed motion and the electromagnetic equations under coordinate transformation needed basic modification. Again Einstein resolved the matter by proposing his theory of special relativity. His brilliant work provided the means for mechanical and electromagnetic equations to be valid in various frames of reference. Along with the new theory of special relativity, Einstein proposed that the speed of light is constant, that any physical phenomenon must be independent of its frame of reference. Any mathematical transformation must preserve this condition, as he proved by applying the Lorentz transformation. A major conclusion of this theory was the transformation of mass and energy through Einstein's famous equation, $E = mc^2$. Although Einstein's theory is based on two postulates, no experimental evidence to date has disproved it.

The creation of quantum mechanics sparked immediate interest in developing new atomic theories and, as a result, various atomic models were proposed. At the same time, work in spectroscopy — discovered by Newton three centuries earlier — was expanding, as new and finer lines found in atomic spectra were raising more questions about the atom. Thus the proposed atomic models had quantitatively to accommodate the occurrence of spectral lines characteristic of any given element. The subsequent refinement of spectroscopy resulted in more precise atomic models.

Among the various atomic models, the Bohr model, later modified by Arnold Sommerfeld, was revolutionary in that it proposed a nucleus with orbiting electrons of discrete energy levels. Based on this model, the radiation line spectra of the element hydrogen could be predicted.

During the early 1920s Louis V. de Broglie proposed a dramatic idea that eventually changed the atomic theory of matter: the dual particle-wave theory, which unites the particle and wave characteristics of light. De Broglie suggested that, in the atomic domain, the particles may behave like waves and waves like particles. He further suggested that optical phenomena can be expressed at certain times by geometrical optics and at other times by wave optics. De Broglie's insight into light, atomic particles, and wave mechanics paved the way for a major breakthrough in quantum mechanics and resolved the 200-year-old conflict between the particle and the wave theories. Aided by the impressive work of several physicists, quantum mechanics was on solid ground by 1930. From 1930 to 1950 the theory evolved at a slower pace, but it was under continual refinement until the theory of quantum electrodynamics was finally established. Today this theory, with some modifications, is the main foundation of atomic physics.

Concurrent with the development of the atomic model was the discovery of the physical phenomena underlying the emission of X rays. This discovery further confirmed the hypothesis of discrete energy levels in the atom.

The developments in atomic theory led to the emergence of solid-state physics, a field in which atomic behavior in metals, semiconductors, and insulators is explored. Knowledge was expanding so rapidly in this field that within a decade a practical solid-state device — the first transistor — was invented. The transistor opened the way for the invention of numerous low-power and compact electronic devices that were essential to the development of computer hardware.

Along with the work being done in physics, optics was branching into yet another field, photography. At the turn of the century technicians had perfected cameras to the point where they could be manufactured and marketed commercially. While discoveries in physics were making room for new, highly technical data-processing machines and computers, the upsurge of interest in photography was laying the groundwork for finely tuned optical apparatuses, new forms of journalism, and experimental work with film. But most important, photography was preparing the way for a cultural revolution.

Interest in photography and film affected all fields of study. Soon after cameras had been accepted as everyday items, the vacuum tube was invented. This was the beginning of what was probably the most far-reaching transformation in history, for it led to the development of radio and television.

The history of the study of light in this period is thus a history of dazzling intellectual achievement and convulsive social change. One can scarcely think of an area of human concern which was not in some way affected by the study of light phenomena between 1900 and 1950.

## 3.2 Revenge of the Particle: The Failure of Classical Physics

By the latter part of the nineteenth century the classical theory of mechanics and electromagnetism was in trouble:

1. The ether hypothesis and the hypothesis regarding propagation of electromagnetic waves in a rest medium had encountered experimental contradictions.

2. The Galilean transformation, upon which Newton's laws of classical mechanics were based, had difficulty in keeping the format of Maxwell's equations invariant under coordinate transformation.
3. The observed spectral radiation of a blackbody could no longer be explained by the Rayleigh-Jeans classical equipartitioning theory.

Thus the second period of scientific advancement in optics (1600-1900), which began with the dramatic invention of the telescope, ended in uncertainty and conflict. The roots of the problem lay in the very foundations of the laws of classical physics, which were formulated during this same period of time.

During the first decade of the twentieth century, three new theories in the field of optics restructured the foundations of classical physics. The first theory, propounded by the German physicist Max Planck[54] (1858-1947), concerned the laws of blackbody radiation. The next two theories, formulated by another German physicist, Albert Einstein[55, 56]* (1879-1955), explored the photoelectric effect and the special theory of relativity.

Although these two scientists were students of the old school of classical physics, their experimental findings could not be explained by its laws. By changing the foundations of classical physics, the special theory of relativity rejected the ether hypothesis and modified the laws of classical mechanics and electromagnetism by way of the Lorentz transformation. The phenomena of blackbody radiation[57] and the photoelectric effect could be explained only by assuming quantization of energy in the atom. This hypothesis was at odds with the assumption of continuous energy in the atom, which prevailed among scientists at that time.

*In 1905 Einstein published three papers. They dealt with the photoelectric effect, the special theory of relativity, and the Brownian thermal-motion analysis.

## 3.3 Correction of a Three-Century-Old Theory: The Special Theory of Relativity

The experimental anomalies between optics and electromagnetic waves which had become obvious by the end of the nineteenth century were in desperate need of a new and radical remedy. This remedy was the special theory of relativity, which was forced upon physics by

1. The results of Maxwell's efforts in unifying optics, electricity, and magnetism;
2. The inadequacy of the laws of classical mechanics in keeping the equation of electromagnetism invariant under the Galilean transformation;
3. The experimental evidence contradicting the existence of a transporting medium, ether.

Hendrik Lorentz made a major contribution to the foundation of the theory of relativity; indeed, his famous transformation could be considered the basis for its formulation. In 1899 Jules-Henri Poincaré was the first to denounce the ether hypothesis and to suggest that the speed of light is the limiting speed for material particles. Between 1900 and 1904 Lorentz and Poincaré showed that the results of the Michelson-Morley experiment were consistent with Maxwell's equations. These results led to "the theory of relativity" by dismissing the idea that bodies have "absolute motion relative to a hypothetical medium" such as ether. Poincaré first named the theory "the principle of relativity" in September 1904.

A year later Albert Einstein forced a crucial shift in the development of the theory of relativity. Convinced of the unacceptability of both the ether hypothesis and the variation of the form of electromagnetic equations under the Galilean transformation, he presented his "special theory of relativity." Einstein based his approach on

1. The constancy of the velocity of light, as all previous measurements had indicated;
2. The axiom that Maxwell's equations must be invariant in form in all inertial frames.

In the following two postulates, Einstein presented the special theory of relativity, which encompassed classical mechanics, electrodynamics, and all natural phenomena, resolving many of their problems:

> *Postulate 1.* The laws of nature, including the laws of mechanics and the laws of electromagnetic phenomena, are the same in all inertial frames of reference and are independent of the translational motion of the system as a whole. More precisely, all inertial frames are completely equivalent for all phenomena.
>
> *Postulate 2.* The speed of light is independent of the motion of the emitting body. Light is propagated in empty space.

The first postulate rejected the idea of ether. "The concept of ether will become superfluous," Einstein said, "as the view here does not require an absolutely stationary space." The experiments conducted by Bradley, Fizeau, Airy, and Michelson-Morley were then quite naturally explained. This postulate is consistent with all our experience in mechanics, where any relative motion between bodies is included. But it required modification under the laws of mechanics for very high speeds.

The second postulate, untested when proposed, was simple in nature and decisively verified only in recent years. Some recent experiments, such as those conducted by CERN* in Geneva[58] in 1964,

*CERN stands for Centre European de Recherches Nucleaires, Geneva, Switzerland.

have conclusively established the validity of Einstein's second postulate.*

The second postulate was a scientific revelation that agitated many physicists. The ether hypothesis of Newton's time had endured for two centuries (or two millennia, if one goes back to the Greeks). The failure of intensive experimentation during the second half of the nineteenth century to confirm the existence of ether and its motion had raised a problem regarding the velocity of both light and ether. Einstein's second postulate helped to solve this problem in a stunning manner.

There is a tendency for new concepts to sink in and be accepted only gradually. But the entire physics community was shocked by Einstein's unorthodox second postulate. The initial difficulty for physicists was that Einstein's second postulate apparently contradicted widely accepted experimental results. Even Lorentz, whose transformation still remains a foundation of relativity, had difficulty accepting the new concept and sorting out the apparent contradiction. Many efforts were made to maintain observed experimental results without using Einstein's second postulate in order to keep intact the old laws of classical physics.

One of these efforts was made by Ritz. He modified Maxwell's equations by saying that the velocity of light was equal to c when measured relative to the source. Thus Ritz could explain all the observed phenomena already discussed, including the Michelson-Morley experiment. His effort proved futile, however, as new experimental evidence using extraterrestrial light sources (such as light emitted from binary stars) was brought forth.

Binary stars are pairs of stars that rotate rapidly around the center of their mass. If light velocity depends on the source motion, then as one star moves toward the earth and the other moves away from it, the rotation of one pair of binary stars should produce an

*The speed of 6 GEV photons produced in the decay of very energetic neutral ions was measured 0.99975 c. Within experimental error it was found that the speed of the photons emitted by the extremely rapidly moving source was equal to c.

unusual orbit. In 1913 the Dutch astronomer Willem de Sitter[59] (1872-1934) showed that the orbital motion of binary stars can be predicted according to Newton's law of motion, as long as the velocity of light emitted from the star is considered constant.

There is still no experimental evidence, even from recent investigations, to cast doubt on the constancy of the speed of light. The velocity of light in a vacuum is independent of the motion of the source and the observer and is a universal constant. Through the Lorentz transformation, the special theory of relativity relates space and time and connects energy, mass, velocity, and momentum for the first time. The ultimate justification of any theory lies in the comparison of its predictions with the results of scientific experiments. No conflicts have yet been found in the special theory of relativity between predictions and the results of experimental investigation.

*Albert Einstein (1879-1955)*

The special theory of relativity has a wide range of application in the field of physics when the interaction of matter is involved. Some important consequences of this theory affect the laws of mechanics. It is remarkable that Maxwell's original equations were comparatively consistent with the theory of relativity and required no modification. Maxwell's equations were formulated in order to represent the results of certain experiments. The equations were repeatedly checked in the laboratory and always found to be correct. In the late nineteenth century several experiments, notably those of Michelson and Morley, were performed to ascertain the accuracy of Maxwell's equations in moving reference frames. The results were consistent and no preferred reference frame was found. The conclusion was that Maxwell's equations are valid in all inertial frames. The special theory of relativity was developed specifically by Einstein to account for this conclusion.

### 3.3.1 The Correction of Common Sense: The Lorentz Transformation

The equations used in relativity to transform space and time variables from one frame, moving with a constant velocity, to another are known as the Lorentz transformation equations:

$$x' = \frac{1}{\sqrt{x - \frac{v^2}{c^2}}}(x - vt)$$

$$y' = y \qquad (3.1)$$

$$z' = z$$

$$t' = \frac{1}{\sqrt{1 - \frac{v^2}{c^2}}}\left(t - \frac{vx}{c^2}\right)$$

When Lorentz proposed these equations his purpose was to represent the earth's velocity with respect to the ether frame. In the special theory of relativity, however, Einstein used a similar form, with v representing the velocity of any inertial frame with respect to any other inertial frame.

When v, the relative velocity of one frame with respect to another frame, is small compared with the velocity of light, the Lorentz transformation reduces to the Galilean transformation. But when v is comparable to the velocity of light, the differences between the Galilean transformation and the Lorentz transformation become significant. When v is larger than c, the spatial coordinate and the time coordinate transform into imaginary coordinates, and the Lorentz transformation loses its value, as expected. The Lorentz transformation has been successfully applied to many problems in quantum mechanics.

## 3.3.2 Contradicting Logic: The Relativistic Velocity Transformation

Assume that the velocity u of a particle, as measured in a frame of reference 0, has $u'_x$, $u'_y$, and $u'_z$ components. Assume also that frame 0′ moves relative to 0 with velocity v. If the velocity vector of the particle, as measured in the moving frame 0′, has the components $u'_x$. $u'_y$, and $u'_z$ , then

$$u'_x = \frac{u_x - v}{1 - \frac{v\,u_x}{c^2}}$$

$$u'_y = \frac{\sqrt{1 - \frac{v^2}{c^2}}\,u_y}{1 - \frac{vu_x}{c^2}} \tag{3.2}$$

$$u'_z = \frac{\sqrt{1 - \frac{v^2}{c^2}}\,u_z}{1 - \frac{vu_x}{c^2}}$$

These equations are called the relativistic velocity transformation. It is impossible to choose u and v so that u′, the magnitude of the velocity measured in the new frame, will be larger than c.

Assume that a particle has a velocity of 0.9 c, as measured in frame O, in the positive x direction, and that a second particle has a velocity of 0.9 c in the negative x direction. According to the Galilean transformation, the speed of one frame relative to the other is 1.8 c, which is impossible. According to the relativistic velocity transformation, equation 1 of (3.2), the relative speed is

$$u' = \frac{0.9\,c - (-0.9\,c)}{1 - \frac{(0.9\,c)^2}{c^2}} = 0.994c \tag{3.3}$$

The velocity transformation shows that c is the limiting velocity for all physical phenomena. As v approaches zero, the Lorentz transformation becomes a Galilean transformation.

### 3.3.3 Dilation of Time: Relativistic Time

Assume two events occur at the same place but are observed in two separate reference frames. If one frame (let's say the second frame) is moving relative to the first with velocity v, then the time interval of an experiment measured in the second frame will be longer than the time interval measured in the first frame by a factor of $(1-v^2/c^2)^{1/2}$. The time interval measured in the frame in which the event occurred in the same place is called proper time. The effect is referred to as time dilation.

### 3.3.4 Contraction of Length: Relativistic Length

If a rod with a length $\ell$ is measured in two frames, with one frame fixed relative to the rod and the other frame moving with a velocity v in the direction parallel to the length of the rod, then the rod is measured as shorter by a factor of $(1 - v^2/c^2)^{1/2}$ in the moving frame, compared with the measurement made in the frame in which the rod is at rest. The length of the rod measured in the frame in which it is at rest is called proper length. The shortening effect of the rod is called the Lorentz contraction.

### 3.3.5 Interconversion of Mass and Energy: Relativistic Mass

As noted earlier, the laws of mechanics need modification for high-speed motion. To keep the form of the equations of classical mechanics invariant in a transformation from one frame to another, some compensating modification is necessary. This change resulted in the creation of a new field of mechanics called "relativistic mechanics." Here Newton's equation of motion and the law of conservation of momentum are maintained, if the mass m is allowed to vary as a function of the velocity; that is, if m = m(v), then (F = dp/dt) between the force, F, and the momentum, p. The mass m(v) is called the relativistic mass. To be consistent with conservation of momentum, the relativistic mass, m(v), must be related to the rest mass, $m_o$, measured when the particle is at rest, by a factor of $(1 - v^2/c^2)^{1/2}$ That is,

$$m(v) = \frac{m_o}{\sqrt{1 - \left(\frac{v}{c}\right)^2}} \qquad (3.4)$$

For example, if a particle is moving at a velocity of v = 0.2c, the relativistic mass is larger than the rest mass by 4 percent.

The relativistic mass increases rapidly as v approaches c. For a finite value of rest mass, $m_o$, the relativistic mass approaches infinity as the particle speed approaches c. Thus it is apparent that the velocity of a particle cannot exceed c.

In many processes in nature the total rest mass of an isolated system changes significantly. It is apparent that the law of conservation of mass found in classical mechanics is no longer valid.

One of the main conclusions of the special theory of relativity is the transformation of mass and energy, according to Einstein's equation, $E = mc^2$. The laws of conservation of mass and conservation of energy, in classical mechanics, do not hold true in relativistic mechanics. But a single law of conservation of the total relativistic energy is valid.

The total relativistic energy E (v) of a particle moving with a velocity of v is

$$E(v) = K(v) + E(o) \qquad (3.5)$$

where K(v) is the kinetic energy of the particle
and E(o) is the intrinsic energy associated with the particle when it is at rest.

The energy E (v) is called the total relativistic energy, and E(o) is called the rest mass energy.

The rest mass energy E(o) of a particle is related to its mass $m_o$ by

$$E(o) = m_o c^2 \tag{3.6}$$

The introduction of rest mass energy, $m_o c^2$, is one of the consequences of relativistic mechanics and represents an addition to classical mechanics. The addition of this new term to the energy balance equations of classical physics does not violate their validity because the total rest mass on each side of the equations is basically the same.

The total relativistic energy E of a particle is $c^2$ times its relativistic mass, that is,

$$E = mc^2 \tag{3.7}$$

where m, the relativistic mass, is

$$m = \frac{m_o}{\sqrt{1 - \left(\frac{v}{c}\right)^2}} \tag{3.8}$$

Equation (3.7) is the famous Einstein energy equation. The total relativistic energy may be written

$$E = K + m_o c^2 \tag{3.9}$$

It is often convenient to have an expression for the total relativistic energy which explicitly includes the momentum p:

$$E^2 = c^2 p^2 + m_o c^4 \qquad (3.10)$$

Under the Lorentz transformation, none of Newton's or Maxwell's equations change form from one frame of reference to another, moving at a velocity constant relative to the first. However, the force in mechanical equations and the electric and magnetic fields in electromagnetic equations do change form under the Lorentz transformation.

*Max Planck (1858-1947)*

After developing the special theory of relativity, Einstein expanded on the general theory of relativity,* in which large-scale gravitational forces are involved. The results of the work conducted from 1905 to 1917 highlight another crucial point in the history of optics:[60] Einstein found that light has mass. The general theory of relativity predicts an exact deflection angle as the light beam bypasses the massive gravitational pull of our own star, the sun. This theory was proven to be true by the experiment made during the eclipse of 1919.

*The special theory of relativity is restricted to reference frames at rest or in motion at constant velocity relative to the fixed stars (inertial frames). The general theory of relativity is an extension to include accelerated frames of reference. A consequence of the general theory of relativity is that light itself interacts gravitationally with objects. Thus a beam of light is deflected as it passes close to so massive an object as the sun.

The special theory of relativity assumes that a photon* has a mass $m = E/c^2$ and is deflected by Newton's gravitational field, like any ordinary projectile. Calculations based on this theory and on Newton's gravitational law show that the deflection angle of a photon just passing by the surface of the sun occurs in approximately 0.87 second.

According to the general theory of relativity, space-time† is curved in the vicinity of a mass such as the sun. The deflection of the light becomes exactly twice the value computed from Newtonian gravity (1.75 seconds). Measurement of starlight near the sun made during the total solar eclipse in 1919 confirmed the results of the general theory of relativity.

Other experiments that verify Einstein's theory include sending a pulsed radar signal from the earth to Venus and Mercury when these planets appear in the vicinity of the sun and measuring the timing of the received signal. The time is measured as the positions of the earth and the planet change relative to the sun. When the beam passes by the surface of the sun, there is a delay in the received signal. The delay, or the increase in travel time, occurs during the trajectory closest to the sun. For example, for Venus, the measured delay near superior conjunction‡ is measured at approximately 220 microseconds. These measurements are consistent with the results from the general theory of relativity.

Einstein's contributions to physics are immeasurable. His two 1905 papers on the special theory of relativity and the photoelectric effect, and his development of the general theory of relativity in 1971,

*A photon is a quantum of electromagnetic energy having both particle and wave behavior. It has no charge or mass but possesses momentum. The term was introduced by C.N. Lewis in 1926.

†Space-time is a term used to express the geometry of the physical universe according to the theory of relativity. Although in the Newtonian physics space and time are two independent entities, it was shown by the theory of relativity that they are intimately intertwined.

‡Conjunction: The meeting of two or more stars or planets in the same degree of the zodiac. The superior conjunction of a planet is its position when it is seen on the side of the sun most distant from the earth.

are among history's most important contributions to science. It is amusing to note that after Einstein published his special theory of relativity, Herman Minkowski[61] (1864-1909), who had been Einstein's professor and who had developed a noteworthy mathematical expression for the transformation of space and time coordinates, said, "I would never have expected such a smart thing from that fellow."

Albert Einstein's contributions, as a scientist, to man's understanding of the universe are unequaled, and he is forever established as a genius of science. But it is important to mention that, throughout his entire life, he was actively involved with human affairs in other spheres. For example, he instigated many crusades for peace and world unity for the good of man. His efforts to bring harmony to the world did not enjoy the success of his scientific achievements. It was thus perhaps appropriate for him to assess his life with the words, "Politics are for the moment. An equation is for eternity."

## 3.4 The Light of the Atom: Beginning of the Quantum Theory

By the latter part of the nineteenth century the classical theory of electromagnetism was in trouble. The problem was that observed spectral radiance of a blackbody could no longer be explained by the Rayleigh-Jeans[48] theory, which had been the reigning theory of radiation at the time. John William Strutt, Baron Rayleigh (1842-1919), and James Hopwood Jeans (1877-1946) were English physicists. They developed what became the Rayleigh-Jeans theory of radiation.

According to their theory, the number of electromagnetic oscillations per unit wavelength interval in the blackbody spectrum approaches infinity as the wavelength approaches zero. In other words, most of the radiation energy eventually resides in oscillations of infinitesimal wavelengths that consist of the ultraviolet or X-ray spectrum of the electromagnetic radiation. The experimental results of the blackbody radiation spectrum, however, did not confirm the Rayleigh-Jeans theory. Although experimental results of the infrared portion of this spectrum closely resembled results in the Rayleigh-

Jeans theory, the ultraviolet portion differed so violently that the departure of the short wavelength radiation from the theoretical prediction of classical theory was called the "ultraviolet catastrophe."

For years the ultraviolet catastrophe created major problems. In 1900, however, Max Planck, who originated the quantum theory, proposed a radiation theory that was a radical departure from the classical theory. When Planck presented his paper, entitled "On the Theory of the Energy Distribution Law of the Normal Spectrum," to the Physical Society of Germany on December 14, 1900, he became the pioneer in a completely new area of physics.

Planck based his theory of radiation on the quantization of electron energy* in an atom. According to his theory, the energy gain or loss of an electron passing from one energy level to another, higher or lower than its initial energy level in an atom, must be discrete and is proportional to the frequency of the radiation emitted:

$$E_f - E_i = h\nu \tag{3.11}$$

where $E_f$ is the electron energy in the final state,

$E_i$ is the electron energy in the initial state,

$\nu$ is the frequency of radiation,

and h is the proportionality constant called Planck's Constant. The magnitude of h was shown to be $6.6 \times 10^{-27}$ erg. sec.

Planck, who was a learned thermodynamicist, initially applied the laws of thermodynamics to a blackbody and described its radiation behavior as thermal oscillation. He connected the thermodynamic energy of the blackbody to its electromagnetic energy through statistical mechanics.† Planck concluded that the probability that a system would have a given energy level decreases exponentially with the magnitude of that energy level:

*A result of quantum mechanics was that the electron energy is discrete rather than continuous. This property of the energy is called quantization of energy.

†Statistical mechanics is the branch of physics that endeavors to explain the macroscopic properties of a system on the basis of the properties of its microscopic constituents.

$$\frac{P_2}{P_1} = e^{-\frac{E_2-E_1}{KT}} \qquad (3.12)$$

where $P_2$ is the probability that the system has an energy level, $E_2$, corresponding to the second state,

$P_1$ is the probability that the system has an energy level, $E_1$, corresponding to the initial state,

k is the Boltzman's Constant,

and T is the absolute temperature.

Equation (3.12) indicates that the probability $P_2$ of a system having an energy level $h\nu$ reduces as the frequency $\nu$ increases. For an infinite frequency, the probability that the system has such an oscillation falls to zero.

Planck's novel theory predicted a substantial reduction in the blackbody spectrum for high-frequency radiation, compared with radiation results calculated by the Rayleigh-Jeans formula. Although his theory was compatible with experimental results, the scientific community, viewing it as a radical departure from the classical approach, preferred to ignore his results.

When Einstein postulated his theory of the photoelectric effect[62] in 1905 (also previously observed by Heinrich Hertz), he proposed that electromagnetic radiation itself be quantized. Einstein suggested that when a quantum of electromagnetic radiation or a corpuscle (later to be called a photon) collides with one electron on the surface of a matter, a minimum of photon energy, called the work function, must be expended to remove the electron from the surface. Einstein postulated that each photon has an energy $h\nu$. If, on the one hand,

$$h\nu > \phi \qquad (3.13)$$

the electron escapes from the surface of the metal and the excess energy appears as the kinetic energy of the electron, that is,

$$h\nu = \phi + \frac{1}{2} m_e v^2 \qquad (3.14)$$

where $m_e$ is electron mass,
and $v$ is electron escape velocity.
If, on the other hand,

$$h\nu < \phi \qquad (3.15)$$

the electron cannot escape.

The ultraviolet catastrophe and the Hertz-Einstein photoelectric effect could be explained only by the hypothesis of quantization of energy. For the second time in five years (1900-1905), the explanation of the experimental results needed the assumption of the quantization of energy. But, because the wave theory was the dominant theory of light (owing to Maxwell's equations), the scientific community remained reluctant to accept the quantization of electromagnetic radiation.

In 1916 Robert A. Millikan[63] (1868-1953), an American physicist, undertook an investigation to disprove Einstein's experimental results of the photoelectric effect, but instead his results verified Einstein's theory. Thus the doors opened for the acceptance of the dualism concept of the wave-particle property of light. The diffraction and interference of light were explained by the wave nature of light, while the photoelectric effect and the emission of blackbody radiation were explained by the quantum theory.

In Planck's radiation law and Einstein's theory of the photoelectric effect, both based on the particle theory of light, the light frequency term $\nu$ is explicitly expressed as the wave nature of light. This implied the inherent connection between the wave and the particle nature of light, which formed the basis of Louis de Broglie's wave-particle dualism theory (see section 3.5).

### 3.4.1 The Atomic Model

The concept of the atom was introduced in the fifth century B.C. by Democritus. The idea comes from the Greek word "atomos," which means the smallest unit of a large set which possesses all the essential characteristics of the set. In this sense the atom is indivisible. Although one can split an atom by separating its proton and electron, its characteristics are destroyed in the process.

The first atomic model was proposed in 1815 by William Prout[64] (1785-1850), an English chemist and physician, who assumed that the hydrogen atom was the basis for all elements. The discovery of the quantization of energy dates to December 14, 1900, when Planck announced his laws of blackbody radiation in a lecture before the Physical Society of Germany in Berlin.

Our knowledge of the structure of atoms was influenced strongly by the investigation of optical spectra of natural elements. About 1860 Gustav Kirchhoff and Robert Bunsen had shown that optical spectra are symbols of the characteristics of elements by their emitting or absorbing light. The earliest absorption lines detected were observed in the solar spectrum. The gases in the sun's atmosphere absorb certain wavelengths of the continuous spectrum emitted by the solar core.

In 1885 Johann Balmer found an orderly principle in atomic spectra of hydrogen atoms, expressed as

$$\lambda = \left(\frac{n_1^2}{n_1^2 - 4}\right) G \qquad (3.16)$$

where $\lambda$ is the wavelength,
$n_1$ is an integer, $n_1 = 3, 4, 5 \ldots$,
and $G$ is an empirical constant.

The creation of the quantum theory prompted intense interest in developing a new atomic theory. The real breakthrough in the atomic model was made by Max Born, Werner Heisenberg, Erwin Schrödinger, Wolfgang Pauli, and Paul Dirac from 1920 through 1930. It had been recognized before 1900 that when a chemical element is either heated or electrically excited, radiation is emitted. Within the spectrum of this radiation a single line or an array of spectral lines, which distinguishes the characteristic of each element, can be observed. Elements contained in stars and in the solar atmosphere can be identified by their spectral lines. Thus any adequate atomic model had to provide accurate predictions regarding the spectral lines characteristic of any element.*

When an atom goes from one discrete excited state to a lower discrete energy state, line radiations are produced. But if the atom's final state of excitation is of higher energy than the initial state, a dark absorption line appears on the spectrum of the transmitted light. The refining of spectroscopy, which resulted in a more accurate measurement of the location of each spectral line, resulted as well in more precise estimations of energy states, thus permitting a more accurate description of the atomic model. The spectral shift of each line and the division of the so-called original lines into two, three, or more lines under magnetic (Zeeman) or electrical (Stark) fields contributed to constant refinement of the atomic model. Both the later extension of spectroscopy to the ultraviolet and infrared regions and the application of the laser to spectroscopy added enormously to the wealth of data that verified the atomic model.

### 3.4.1.1 The Thomson Model

Joseph J. Thomson[65] (1856-1940), an English physicist, during an 1895 investigation of cathode rays, discovered the electron.

*The most important source for verification of any conceivable atomic model at that time was the observed optical spectral lines of the atom.

*Joseph J. Thomson (1856-1940)*

This prompted him to suggest an early description of atomic structure. He postulated that an atom contains diffused charge clouds rather than a concentrated nuclear charge. In the Thomson model, the atom consists of a spherically shaped positive charge on which negatively charged electrons are uniformly located. The radius of each of the positive charges is assumed to be about $10^{-10}$ meters. In the lowest energy state, electrons are fixed at their equilibrium positions. When the atom is excited, as by increasing its temperature, the electrons vibrate around these equilibrium positions. The classical theory of electromagnetism predicts that accelerated charges emit radiation and that the vibrating electron is constantly emitting energy in the form of radiation, thus violating the conservation of energy law. The quantitative description of this radiation was not confirmed by experimental data at the time.

### 3.4.1.2 The Rutherford Model

In 1911 Ernest Rutherford[66] (1871-1937), an English physicist and a student of Thomson's, put forth another model of the atom.

He proposed that an atom is formed by a central nucleus of small dimensions ($10^{-13}$ to $10^{-12}$ cm) around which a certain number of electrons revolve in a circular orbit. The radius of the orbit of the electron is about $10^{-8}$ cm; the atom therefore is composed mostly of empty space. Almost all the mass of the atom is concentrated in the nucleus, which carries a positive electrical charge. This charge is exactly equal to the negative charge of the electrons. Thus if n is the number of electrons in the atom and e is the magnitude of the electron charge, the charge of the nucleus is +ne. The Rutherford model actually resembles a miniature solar system in which gravitational force is replaced by electrical force.

*Ernest Rutherford (1871-1937)*

Classical electromagnetic theory requires that an accelerated charged particle emit electromagnetic radiation. In its orbital motion around the nucleus, the electron is continuously changing direction. It is therefore under an accelerating force and should emit a continuous radiation. But when the electron emits radiation, it loses energy; thus its orbital radius would decrease until it collapses with the nucleus. Such a collapse, of course, does not occur, nor does continuous emission of radiation take place. Excited atoms emit radiation only at selected discrete wavelengths called line spectra.

### 3.4.1.3 The Bohr Model

In 1913 Niels Bohr[67] (1885-1962), a Danish physicist, analyzed the contradictions between the classical theory of radiation predicting a continuous emission of radiation and the experimental results for the hydrogen atom which indicated only discrete spectral lines. He postulated a new atomic model for hydrogen, basing his approach on the fact that any atomic model must adequately explain the origin of the line spectra. Bohr proposed that the hydrogen atom has the following charateristics:

*Niels Bohr (1885-1962)*

1. The single electron in the hydrogen atom moves around a stationary nucleus in circular orbits. In these orbits the electrostatic attraction between the negatively charged electron and the positively charged nucleus is exactly equal to the centripetal force needed for a circular orbit.

2. The energy of the electron in its orbit is constant. The electron remains in its orbit unless it absorbs or emits an exact quantum of energy.

3. Only certain orbits are allowed — namely, those for which the magnitude of angular momentum $\ell$ of the electron, defined as $m_e vr$, is an integral multiple of $\hbar = h / 2\pi$ :

$$\ell = m_e vr = n\hbar = n\frac{h}{2\pi} \qquad (3.17)$$

where $m_e$ is the linear velocity,
$v$ is the radius of the orbit,
$r$ is the electron mass,
and $n$ is an integer and is called the principal quantum number.

When the electron absorbs photons with an amount of energy corresponding exactly to a particular wavelength, it can move from a lower energy state to a higher one, as established by the equation

$$E_2 - E_1 = h\nu \qquad (3.18)$$

Conversely, when an electron moves from a higher energy state to a lower one, a photon will be emitted.

According to Bohr, if the atom makes a transition from a state with energy $E_1$ to another state of energy $E_2$, the frequency of the radiation emitted is

$$\nu = \frac{2\pi^2 mz^2 e^4}{h^2}\left(\frac{1}{n_2^2} - \frac{1}{n_1^2}\right) \qquad (3.19)$$

where $n_1$ and $n_2$ are the quantum numbers corresponding to the energy states $E_1$ and $E_2$, respectively,
and $z$ is the atomic number.

In terms of the wave number* $\bar{\nu} = \frac{\nu}{c}$, equation 3.19 becomes:

$$\bar{\nu} = \frac{2\pi^2 mz^2 e^4}{ch^2}\left(\frac{1}{n_2^2} - \frac{1}{n_1^2}\right) \qquad (3.20)$$

This last postulate is in agreement with Einstein's quantum hypothe-

*Wave number, $\bar{\nu}$, is the reciprocal of wavelength $\lambda$, i.e. $\bar{\nu} = \frac{1}{\lambda}$.

sis for electromagnetic radiation.

The value of Bohr's model was in the precision with which it predicted the spectral lines for hydrogen. The allowed energy state $E_2$ is found to be

$$E_i = \frac{-R^2}{n_i^2} \tag{3.21}$$

where

$$R = 2\pi^2 m_e \frac{e^4}{h^2} = 2m_c \frac{e^4}{\hbar^2} \tag{3.22}$$

R is the Rydberg constant[68].

The energy $\epsilon$ emitted in the transition of a single electron from an energy state $E_i$ to an energy state $E_j$ is therefore

$$\epsilon = h\nu = E_i - E_j = R\left(\frac{1}{n_j^2} - \frac{1}{n_i^2}\right) \tag{3.23}$$

The theoretical model of this equation, particularly when corrected for the finite mass of the nucleus, closely paralleled Johannes Rydberg's experimental results.

Bohr's atomic model explained many physical phenomena associated with optical radiation. Despite some shortcomings (see the next section), his theory marked a giant step forward in physical science. His theoretical prediction, combined with experimental evaluation, initiated a period of rapid growth in the comprehension of the microscopic structure of matter.

### 3.4.1.4 Shortcomings of Bohr's Model and Sommerfeld's Contribution (Elliptical Orbit)

Bohr's model, though undoubtedly a valuable contribution to the development of the atomic model, failed in three respects:

1. When the single lines in Bohr's model were viewed with a higher-precision spectrometer, they indicated multiple spectral lines which were components of the original single lines. Bohr's model could not explain these multiple spectral lines.
2. It could not explain the line intensities in the hydrogen spectrum.*
3. It failed to interpret the energies and spectra of more complex atoms.

In 1916 Arnold Sommerfeld[69] (1868-1951), a German physicist, revised Bohr's model to incorporate elliptical orbits to replace circular orbits. Although Sommerfeld's approach is now obsolete, his line of reasoning has historical value. Sommerfeld applied Kepler's laws and classical mechanics to conclude that, in addition to circular orbits, elliptical orbits of the same energies are possible. He then introduced a second quantum number k, so that $k \leqslant n$. The two quantum numbers n and k correspond to the major and minor axes of the ellipse:

$$\frac{n}{k} = \frac{a}{b} \qquad (3.24)$$

where a is the length of the major axis,
and b is the length of the minor axis.
When k = n, the orbit becomes Bohr's circular orbit.

Sommerfeld identified k, a measure of the eccentricity of the

*Line spectrum refers to the discontinuous spectrum emitted from atoms, ions, and certain molecules, when they are in an excited gaseous phase.

ellipse, where an angular momentum of the atom will be kh. The hypothesis of the elliptical orbit, however, had no effect on the energy level of the atom and therefore provided no explanation for the fine structure of spectral lines.

Later on, Sommerfeld's second quantum number, k, was changed in the language of quantum theory to the angular momentum quantum number $\ell$, where $\ell = k - 1$. Sommerfeld then successfully applied the effects of relativity to moving electrons in an elliptical orbit and obtained a finer structure of the line spectra. He reasoned that when an electron accelerates as it approaches the nucleus, it requires an acceleration similar to that in Kepler's planetary motion equation in order to sweep an equal area between the nucleus and its orbit in unit time. When the orbital radius is smaller, the electron must move faster. According to the special theory of relativity, when an electron moves faster, then its mass increases and its energy reduces. For the minor axis of the ellipse, this relativistic correction becomes more significant. Sommerfeld's relativistic calculation resulted in an electron energy $E_{n,k}$:

$$E_{n,k} = E_{Bohr}\left[1 + \frac{\alpha^2}{n^2}\left(\frac{n}{k} - \frac{3}{4}\right)\right] + \text{higher order corrections} \qquad (3.25)$$

$$\text{where } E_{Bohr} = \frac{Rhcz^2}{n^2} \qquad (3.26)$$

z is the nuclear charge number,

and $\alpha$ $= 7.29735 \times 10^{-3}$ and is called the Sommerfeld fine structure constant.

But even with this refinement the Bohr-Sommerfeld model still could not explain the subdivisions of one line into several finer lines. The detection of the normal Zeeman effect,[70] by which each spectral line of the hydrogen atom splits into three components while under the influence of a magnetic field, necessitated the introduction of a third quantum number. The interaction of an external magnetic field of an equivalent magnet produced by the motion of the electron in its orbit splits the spectral line into three components. In this situation, the component of the angular momentum in the direction of the magnetic field is $mh/2\pi$, where m is a positive or negative integer less than 1. These results indicate that an atom can have only a specific orientation in space. If $\ell = 1$, m can be only -1, 0, or +1. That is, under the influence of the magnetic field, atoms can take three positions, in which the angular momentum is parallel, perpendicular, or antiparallel to the magnetic field.

*Wolfgang Pauli (1900-1958)*

Despite several important modifications of the Bohr-Sommerfeld model in the years following 1916, the revised model was not able successfully to interpret the line spectra of atoms containing more than one electron. The intensities of some finer line structures were still completely unaccounted for.

The problems associated with the fine structure of the spectra of the extended Bohr model were first overcome by an Austrian-born American physicist, Wolfgang Pauli[71](1900-1958), in 1925. Pauli suggested that a fourth quantum number be associated with the spin of an electron* of magnitude $1/2\hbar$. This contrasted with Bohr's orbital angular momentum, which was a whole multiple of $\hbar$. According to quantum rules about angular momentum, the spin can be aligned only along or opposite to a specified direction, giving the required two values.

Pauli also made the assumption that no two electrons in an atom can have the same set of values for the four quantum numbers. The Pauli Exclusion Principle was subsequently applied to any particle having a spin that is an odd multiple of 1/2. From this principle, Pauli concluded that the set of four quantum numbers (principal, angular, magnetic, and spin) uniquely define the energy state of an electron in a stable atom. These restrictions helped to build the quantum model of complex atoms.

## 3.5 Healing the Wounds: Wave Mechanics

By 1920 the atomic model of Bohr and its extension by Sommerfeld and other scientists could explain some but not all experimental results. Furthermore, the equations borrowed from classical mechanics, such as the Keplerian orbital motion equation, were not consistent with the hypothesis of the quantization of energy. A completely self-sufficient quantum mechanical theory explaining the atomic model was necessary. Nevertheless, the experiments from 1900 to 1920 were valuable in showing that the particle theory of light is valid.

Once again, confusion reigned regarding the nature of light. At this time the phenomena of diffraction and interference were

*Spin of an electron refers to that property of an electron which gives rise to its angular momentum about an axis within the electron. This angular momentum should be distinguished from the orbital angular momentum.

explained by the wave nature of light, while those of the photoelectric effect and blackbody radiation were explained by the particle theory of light. As both these properties of light had been successfully tested in numerous experiments, the scientific community of the 1920s seemed ready to resolve the conflict.[73] The solid experimental evidence of the particle and the wave nature of light prepared the stage for a unified theory that would explain them both.

Although few developments in the early twenties had a lasting effect, accomplishments in the field of physics were prodigious. Efforts were made, not to explain the wave or particle nature of light, but to develop a new theory that would unify the two properties.

In 1923, influenced by the prevailing spirit of inquiry, the French physicist Louis V. de Broglie[74] (1892-1960), a man of insight and imagination, proposed a unified theory for photon and atomic particles. He observed a fundamental difference between particles and waves in general. The energy of a particle is concentrated in a small area of space confined to the particle, whereas waves distribute their energy in many directions and expand as a function of time.

Furthermore, the trajectory of a particle follows the law of classical mechan-

*Louis V. de Broglie (1892-1960)*

ics based on Newtonian laws, whereas waves undergo diffraction and interference that cannot be explained by the Newtonian mechanics of a trajectory. De Broglie, observing that these differences are obvious in the domain of classical physics (which primarily treats the macroscopic behavior of matter), believed that in the atomic domain particles may behave like waves and waves may behave like particles.

De Broglie noted that optical phenomena can be expressed at certain times by geometrical optics and at other times by wave optics. For long distances and large objects, the rules of geometrical optics describe the process in which rays at certain trajectories form an image.[37] But in this case these rules resemble the evaluation of particle trajectory in classical mechanics. The refraction of light as it passes from one medium to another is analogous to the change in the trajectory of a particle under the influence of different forces. For small objects, however, or when the diffraction and interference of light are considered, then the resulting phenomena can be explained only by the wave properties of light.

De Broglie was influenced by the work of the Irish mathematician William Hamilton (1805-1865), who in 1835 had written an equation of motion for a material point in a field of force which is similar to the equation of a light ray in a medium of varying refractive index. This analogy inspired de Broglie to unite the particle and wave characteristics of light. He noted that the laws of geometrical optics can be derived as an approximation to the limits of the laws governing wave optics. By analyzing the comparison between waves and particles just described, de Broglie suggested his hypothesis:

> *Associated with any material particle moving with momentum P, there is a wave of wavelength $\lambda$ related to P according to the equation*

$$\lambda = \frac{h}{P} \qquad (3.27)$$

where h is Planck's Constant."

De Broglie observed that the energy $h\nu$ of a light corpuscle contains the frequency $\nu$. He concluded that in the case of light, the ideas of corpuscle and of periodicity (a nature of wave) are simultaneously present. He stated, furthermore, that the motion of electrons in the atom is limited to orbits whose positions are defined by the principal quanta. The only phenomena in physics that contain integer numbers in their solutions are those of the wave nature of vibration and interference. Therefore, de Broglie extended his particle-wave dualism theory of light to electrons, and hence to all particles.

According to de Broglie, associated with a particle such as an electron is a system of "matter waves" or "particle waves"that have slightly different velocities of propagation. At regular intervals along the path of propagation, the waves combine to form a crest. This crest disappears at one point and reappears at another point in a later instant. The velocity of the crest, or "group velocity," is different from the velocities of the individual waves that combine to form it. De Broglie identified the group velocity as the velocity of the electron. The distance between successive crests in the de Broglie wavelength of the matter waves is*

$$\lambda = \frac{h}{mv} \qquad (3.28)$$

*The application of mechanics to the light wave or particle wave associated with the quantum theory is sometimes called "wave mechanics."

where h is Planck's Constant,
and mv is the momentum of the particle.
It is important to note that this equation supports Bohr's earlier model of the hydrogen atom.

In 1927 American physicists Clinton J. Davisson (1881-1958) and Lester H. Germer[75] (1896-1971), working with slow electrons (59 ev), and English physicist George Thomson[64] (1892-1975), working with fast electrons, showed that a beam of electrons having a common velocity exhibits diffraction (a wave property) when impinging on a crystal lattice. The wavelength calculated from the energy and the angle of deviation definitely agree with the value of $\lambda$ predicted by the de Broglie equation. Much later, in 1961, C. Jonsson demonstrated the wave nature of electrons again by duplicating Thomas Young's double-slit experiment with an accelerated beam of electrons. The interference patterns for accelerated electrons and light were strikingly similar.

Applying the results of the special theory of relativity, de Broglie refined his argument, proposing that if a particle of rest mass $m_o$ and, according to relativistic dynamics, an energy $mc^2$ has associated with it a frequency $\nu_\sigma$, then

$$\nu_o = \frac{E}{h} = \frac{mc^2}{h} \qquad (3.29)$$

If the particle moves relative to an observer with a velocity v, the observer will see the frequency shifted by the relativistic time dilation, defined by the theory of relativity to a value of

$$\nu = \nu_o \sqrt{1 - \frac{v^2}{c^2}} \qquad (3.30)$$

But the energy of the moving particle is such that it corresponds to a frequency:

$$\nu_1 = \frac{\nu}{\sqrt{1 - \left(\frac{v}{c}\right)^2}} \qquad (3.31)$$

De Broglie proved that these two waves always move in phase, since $\nu$ from equation (3.30) is at the same speed as the particle in the second equation (3.31) at a different speed v, which de Broglie called the phase wave. He compared this phase-wave velocity to the velocity of the ordinary wave motion in a dispersive medium. His analogy was that when a group of waves with slightly different frequencies travel at slightly different speeds, they produce, by interference, a group wave that travels at a speed significantly different from that of each wave. It is this group velocity that represents the speed of transport of energy by the waves. Using this parallel, de Broglie deduced the wavelength of these matter waves as

*Arthur H. Compton (1892-1962)*

$$\lambda = \frac{h}{P} \tag{3.32}$$

which has already been discussed.

As de Broglie's hypothesis is not deduced logically, it would seem that quantum physics based on this hypothesis is not built on a logical foundation. Although that is still true today, the important fact is that the development of the quantum theory followed de Broglie's work and accorded completely with experimental results. The mathematical formulas that are the building blocks of the quantum theory enable us to compute the detailed properties of atoms and their nuclei with incredible precision. The results of this mathematical finding, as tested by experiment, have consistently been found to be accurate.

*Werner Heisenberg (1901-1976)*

De Broglie's hypothesis was a source of inspiration for many scientists of the 1920s. Through his insight into the wave-particle nature of light and atomic particles and wave mechanics, he opened the door for a major breakthrough — the formulation of the quantum theory. According to Emilio Segrè[62], an American physicist born in Italy in 1905, "the discovery of the dual aspect of matter is the greatest conquest of the century."

In 1923 the great American physicist, Arthur H. Compton[76] (1892-1962), discovered that if a beam of X rays impinges on a certain material, the X rays will be scattered by the loosely bound electrons of atoms in that material, and the frequency of the X rays will shift in accordance with the principle of conservation of energy and momentum. Compton's experiment verified the hypothesis that X rays carry light quanta of specific momentum.

*Erwin Schrödinger (1887-1961)*

After the discovery of Compton's effect, Adolf Smekal[57] in 1923 and Werner Heisenberg (1901-1976) in 1925 predicted that an effect similar to Compton's would be found for visible light. Chandrasekhara V. Raman (1888-1970), an Indian physicist, had already been investigating this phenomenon. In 1928 he published his conclusion that when an intense and monochromatic light with a frequency $\nu$ is incident at a transparent substance, the spectrum of the scattered light in addition to the original frequency (Rayleigh scattering) also has weaker radiation of frequency $\nu \pm \nu'$. The scattering of these sidebands is called Raman scattering. The Raman scattering has weak lines on each side of the Rayleigh line. If the frequency of the incident light is changed, the scattered radiation again appears in both sides of the incident light with the same frequency difference.

## 3.6 Wave Theory of Atomic Particles

In 1926 Erwin Schrödinger[77,78] (1887-1961), a German physicist deeply influenced by de Broglie's work, incorporated Hamilton's formulation of the wave equation and developed a true wave theory for atomic particles. Schrödinger's famous wave differential equation contains the energy of the particles as a parameter:

$$\nabla^2 \psi(x, y, z) = \frac{-8\pi^2 m}{h^2} \left[ E - U(x, y, z) \right] \psi(x, y, z) \tag{3.33}$$

where $\nabla$ is the Laplacian operator,
$E$ is the energy of the system,
and $U$ is the potential energy of the system.

In this equation, $\psi$ is the wave function, which has a well-defined wavelength and phase velocity.

The initial problem with the Schrödinger equation was the interpretation of the nature of the wave function $\psi$. This wave function can represent neither the wave amplitude, as it might in the case of water, nor pressure, as in the case of sound. However, $|\psi(\vec{x})|^2$ gives the probability of finding the particle (such as an electron) at a given volume element dv:

$$|\psi(\vec{x})|^2 \, dv = P(\vec{x}) \, dv \tag{3.34}$$

where $P(\vec{x})$ is the probability density function and dv is centered at $\vec{x}$.

Schrödinger initially interpreted $\psi$ as an electron charge diffused in a cloudlike manner and assumed that $|\psi(\vec{x})|^2$ is a percentage of the electron charge that is located at point $\vec{x}$. Later, it was Max Born (1882-1970), a German physicist, who, getting a clue from his own treatment of collision theory*, defined $|\psi(\vec{x})|^2$ accurately as the probability of finding a discrete electron charge at a volume element dv.

When a particle is bound to a confined area, $|\psi(\vec{x})|^2$ gives the probability of finding that particle within a dimension dv centered at $\vec{x}$. Another conclusion from this equation is that if the particle is confined to only one dimension of the length L, we cannot know exactly where the particle is located on that length. In other words, the position of the particle at $\vec{x}$ is uncertain. This uncertainty as to the position of a particle and the assignment of a probability to its presence anywhere within the region of confinement were factors in the development of probabilistic physics.

Physics before Schrödinger was primarily deterministic; all phenomena had to have a clear cause-and-effect relationship. The macroscopic phenomena of nature observed in everyday life always had to be confirmed in this way. When a well-defined force is applied, a well-defined motion takes place. Newton, Maxwell, and Einstein all subscribed to deterministic physics. After the 1920s, however, physics underwent a fundamental change, and the new discipline of probabilistic physics was in the making.

When the microscopic structure of matter(i.e., of atomic particles) is under investigation, an entirely different physical world is manifested. The mere observation of an atomic particle, which requires that photons impinge on it, disturbs the state of the particle. And, since these photons are not subject to experimental controls, we cannot predict or correct their effects. Even when atomic particles can be observed, the disturbance produced by the incident photon imposes limitations on the verification of theories.

Because of the difficulty of observing and verifying particle parameters (including coordinate, momentum, etc., in the atomic

*Collision theory is the theory that describes any interaction between particles, aggregates of particles, or rigid bodies in which they come near enough to exert a mutual influence, generally with exchange of energy. The term collision, as used in physics, does not necessarily imply actual contact.

scale), the quantum-mechanical* solutions are probabilistic. This required physics to depart from its traditional deterministic conception of physical theory. And, since that conception was a very fundamental one, its rejection was sure to be criticized. The transition, from the deterministic theory of earlier physics to the probabilistic theory of modern physics, was in fact a painful one for many physicists of the 1920s, and it met considerable resistance. Proposals such as Heisenberg's uncertainty principle (discussed later in this section) were then not easy to comprehend, Still, since the deductions made from the quantum theory have been consistent with experimental results, resistance has largely disappeared.

The wave equation (3.33) has solutions for only certain values of E. These values are called energy eigenvalues, and their ensembles form a spectrum. In the case of hydrogen, a discrete spectrum results from the wave equation which is identical with Bohr's energy levels. The energy quantization thus happens automatically.

Acceptable solutions to equation (3.33) can be obtained only for certain values of energy. Using this equation, Schrödinger found the values of the energy comparable to Bohr's values:

$$E = \frac{2\pi m z^2 e^4}{n^2 h^2} \qquad (3.35)$$

The solution of the Schrödinger wave equation revealed some of the deficiencies in the Bohr-Sommerfeld orbital model. An outstanding example was the problem of angular momentum. Before Schrödinger, many formulas involving angular momentum had to be artificially manipulated in order to give quantitatively correct answers. The problem was the orbital quantum numbr k, which was variably replaced by one of these terms, $\sqrt{k(k+1)}$, k, or by 1/2. But in Schrödinger's equation, the orbital angular momentum $\ell$ is

*Quantum mechanics is the modern theory of matter, of electromagnetic radiation, and of the interaction between matter and radiation; it is also the mechanics of phenomena to which this theory may be applied.

applied in the form of $\sqrt{\ell(\ell+1)\hbar^2}$, with corresponding values ranging from 0 to n - 1. Here no further artificial manipulation is needed.

The old orbital model was sufficient to explain the simplest hydrogen atom, but it failed beyond that point. It could not provide an arrangement of orbits which was more than marginally stable, nor could it account for the nature of spectra. With the inclusion of spin and the requirement of asymmetry, the Schrödinger equation gave a description that was completely satisfactory. With this solution, Schrödinger's theory scored a clear triumph over the orbital model. Schrödinger's formulation also expressed Pauli's Exclusion Principle mathematically in terms of the properties of the function obtained as a solution of the equation.

*Paul Dirac (1902-1984)*

In 1926 Paul Dirac[79] (1902-1984), an English physicist, independently showed that Pauli's principle is equivalent to the requirement that electrons must be described by an antisymmetric $\psi$, defined as follows: all electrons in an atom are identical; no physical differences can result by interchanging their positions; therefore the probability density $|\psi(\vec{x})|^2$ must remain the same. According to equation (3.34), the wave function $\psi$ must become $\pm\psi$ if the two electron positions are interchanged. If $\psi \rightarrow -\psi$, the wave function is symmetric; if $\psi \rightarrow +\psi$, it is antisymmetric.

Dirac also substantiated some of the more perplexing consequences of Schrödinger's equation. According to Schrödinger, when the probability of the existence of a particle between two walls of a box with finite dimensions is calculated, some nodes indicating zero probability appear in the box. The probability of the existence of the particle at those nodes is zero. This raised the question of how a particle can have a nonzero probability of being on one side of the node, a nonzero probability of being on the other side, but a zero probability of occurring in the node. The answer came with the more accurate relativistic version of the Schrödinger equation. Applying relativity, Dirac showed that the probability of a particle occurring in the Schrödinger nodes is not really zero. Though very small, the probability is in fact nonzero. It was also found in the late 1950s that Pauli's Exclusion Principle can be derived from relativistic quantum theory. For years before that date, Pauli's principle was always an added postulate.

Initially, Schrödinger took into account relativistic mechanics for the motion of electrons, but he was disappointed to find that his results did not agree with experimental observations. The discrepancy was owing exclusively to his not accounting for the spin of the electron, which was recognized in 1925 by C. L. Goudsmit and George Uhlenbeck.[80] Schrödinger noticed, however, that if he treated the electron nonrelativistically, his method produced results in agreement with observations in the relativistic approximation. It was therefore in this form that he published his work in 1926.

Schrödinger's wave theory, later substantially modified by other scientists, is the main support of the quantum theory. It is the key formula in atomic and particle theory, and it has been one of the most important mathematical breakthroughs in physics in the twentieth century, ranking with Newton's eighteenth-century formulation of gravitation and Maxwell's nineteenth-century equations of electromagnetism.

It is interesting to note that, although the three sets of equations for gravity, electromagnetism, and the quantum theory were derived from different fields, not one of them was derived from fundamental principles. Their main virtue is that their mathematical conclusions are consistent with experimental results.

The work of another German physicist, Werner Heisenberg[81,82], parallels Schrödinger's investigations. In 1925 Heisenberg approached the development of an atomic model by applying matrix algebra. His model was completely substantiated by experimental results. In 1926 Schrödinger proved that Heisenberg's results were in agreement with his own wave-mechanics approach.

One of Heisenberg's previous successes had been his development of the theory of the helium atom. There are two families of energy levels* in helium, and it is impossible to pass from one to the other by the emission of radiation. In one system the spins of electrons are parallel to each other; in the other they are antiparallel. Heisenberg correctly calculated wave functions in the helium atom according to the laws of quantum mechanics.

In Heisenberg's application of matrix algebra to the development of an atomic model, he was assisted by two other German physicists, Max Born[83] and Pascual Jordan (1902-1980). Schrödinger's wave-mechanics approach, aided by Max Born's interpretation of $\psi$ , was more widely used than Heisenberg's approach because most scientists at the time were not familiar with matrix algebra. Schrödinger's approach, however abstract, gave an almost pictorial representation of an atom.

According to the uncertainty principle, an important result of Heisenberg's approach, it is impossible to determine simultaneously and with complete accuracy the position and the momentum of a particle, that is,

*Atomic energy levels refer to the definite energy values which correspond to the state of excitation of an atom.

$$\Delta x \cdot \Delta P = nh \qquad (3.36)$$

where $\Delta x$ is uncertainty in x position of the particle and $\Delta P$ is uncertainty in momentum of the particle. This principle has significance on an atomic scale but plays no role in ordinary experience because of the very small value of Planck's constant h.

Einstein vigorously resisted the uncertainty principle and tried on numerous occasions to formulate an argument to refute it. But each time a counterargument, usually presented by Bohr, defeated Einstein's argument. Yet Einstein remained convinced that the deterministic approach was preferable to the probabilistic one. Ending his futile attempts, Einstein wrote to Bohr: "God does not play dice with the universe." The scientific debate and the consequent letters between Einstein and other scientists, in particular Bohr, regarding the uncertainty principle are some of the most inspiring and entertaining technical arguments in the history of science. Max Jammer in his excellent book, *The Philosophy of Quantum Mechanics,* has quoted many of these letters.

Despite the recognized power of the quantum theory, the details of the exact structure of atoms were not known. There simply were no techniques for solving any but the simplest partial differential equations. Even today, although Schrödinger's equation for an atom containing only one electron can be solved exactly, approximation methods must be used for an atom with two or more electrons. Molecules or aggregates of atoms in bulk matter are even more complex.

## 3.7 Perfecting the Theory: Relativistic Quantum Mechanics

Defects in Schrödinger's approach surfaced a few years after publication of his results. First, the spin of the electron had to be included on an ad hoc basis. Second, the equation was not in harmony with the theory of relativity. Numerous efforts (including those of Schrödinger himself) to remedy the second defect were unsuccessful. The difficulty was that the relativistic expression for the energy E for a free particle is

$$E = \pm \left[(c\,P)^2 + (mc^2)^2\right]^{1/2} \qquad (3.37)$$

where m is the mass of the particle,
p is the momentum of the particle,
and c is the velocity of light.

This equation cannot be used directly, as the square root has no clear meaning once the transition is made to the wave equation operators of either Schrödinger's derivatives or Heisenberg's matrix.

In 1928 Dirac[84,85] solved the problem by establishing the formula

$$E = c(\alpha_x P_x + \alpha_y P_y + \alpha_z P_z) + \beta mc^2 \qquad (3.38)$$

where $P_x$, $P_y$, and $P_z$ are the three mutually perpendicular components of momentum satisfying

$$P_x^2 + P_y^2 + P_z^2 = P^2 \qquad (3.39)$$

and $\alpha_x$, $\alpha_y$, $\alpha_z$, and $\beta$ are determined by the requirements that they be independent of position and time, such that

$$E^2 = (c\,p)^2 + (mc^2)^2 \qquad (3.40)$$

A combination of $\alpha$'s and $\beta$'s, compatible with the energy equation (3.36), must be added to the orbital angular momentum $\ell$. This combination must represent an angular momentum that does not depend on the motion of the particle and is therefore an intrinsic angular momentum or spin. When the effect of an electromagnetic field is included in the equation, the spin is found to be associated with a magnetic moment of the exact magnitude proposed by Uhlenbeck and Goudsmit.

Thus Dirac's proposal solved two problems at once. It also implied the existence of a particle similar to electrons but having a positive charge. The introduction of this new particle faced resistance until 1932, when Carl D. Anderson (b. 1905), an American physicist, discovered it in cosmic radiation and called it a positron. Later, these particles were produced in some radioactive processes.

A whole series of discoveries were made in 1932. The neutron was discovered by James Chadwick[48] (1871-1944), an English physicist, and experimental investigations were soon conducted to demonstrate the specific nuclear forces between neutron-neutron, proton-proton, and proton-neutron. Next came the discovery of deuterium, a hydrogen isotope, by Harold Urey[75] (1893-1981), an American physicist. Discoveries relating to beta-ray theory and artificial radioactivity were made in the same year. (See section 3.8.)

Dirac's relativistic version of the wave equation enjoyed significant success and its applications to the electron were consistent with experimental evidence. Dirac found a new level of freedom for the spin of an electron with a value of $h/2\pi$.

Even before he developed the relativistic theory of electrons, Dirac had devised a method for bringing electrodynamics into the realm of quantum theory. After he developed the relativistic theory of the electron, there was a feeling that a turning point in the quantum theory had been reached. The development of quantum electrodynamics,* however, was delayed until the late 1940s, partly because of

*Quantum electrodynamics is the study of the properties of electromagnetic radiation and its interaction with electrically charged matter — in particular, with atoms and their constituent electrons. The fundamental equations governing quantum electrodynamics are believed to encompass all of atomic physics, chemistry, properties of bulk matter, and classical electromagnetic theory.

experimental revelations discussed below.

According to Dirac's relativistic theory, the lowest energy level of a hydrogen atom must have one energy state, and the next higher energy level should have two different energy states. During the 1930s it was found that there were some energy shifts from the above predicted energy levels. In order to explain these results it was assumed that the electrons in an atom are continually emitting and reabsorbing the light quanta, a process in which energy conservation is violated. Dirac's relativistic theory could not fully explain this phenomenon. Dirac's theory also ascribes a magnetic moment of definite magnitude to the electron. But a careful measurement in 1941 by the American physicist Palykamp Kush[86] (b. 1911) showed that the two energy states differ from Dirac's prediction by approximately 0.1 percent.

*Richard P. Feynman (b. 1918)*

In 1947 the American physicists Willis Lamb, Jr. (b. 1913) and Robert Rutherford[87] detected, with the use of high-frequency microwave spectroscopy, still finer resolution lines, referred to as the hyperfine structure in the hydrogen spectrum. These lines were a representation of a fine energy shift known as the Lamb shift. The

*Julian S. Schwinger (b. 1919)*

Lamb shift was very important in the development of quantum electrodynamics. According to the previous theory, the hydrogen atom could exist in two states of the same energy. When Lamb passed microwaves through hydrogen atoms in one state, however, they were converted to the other state. The absorption of the microwave energy proved that the two states did not contain the same energy. The energy difference, though miniscule, was important enough to require a change in the theory.

In 1949, after the discovery of the Lamb shift, several researchers independently and almost simultaneously advanced relativistic quantum electrodynamics, eliminating some of the difficulties previously encountered. They explained that the source of the Lamb shift is the interaction between the electron and the electromagnetic field. At this time, two American physicists, Richard P. Feynman[88] (b. 1918) and Julian S. Schwinger[89] (b. 1919), and the Japanese physicist Sin I. Tomanaga[90] (1906-1979), simultaneously applied quantum mechanics to the electromagnetic field to formulate the theory of quantum electrodynamics.

The atomic model that was obtained as the result of these advances had provided insights into the relation between the partly filled outer valence-band electrons and the chemical properties of matter.

## 3.8 The Alchemist's Dream: Radioactivity

### 3.8.1. Historical Development

*Henri Becquerel (1852-1908)*

Radioactivity was discovered by Henri Becquerel (1852-1908), a French mathematician, in 1896, shortly after the discovery of X rays by Wilhelm Röntgen in 1895. Becquerel's main interest, stirred by a generations-old family preoccupation, was the phosphorescence and fluorescence of various materials. Upon learning of Röntgen's discovery of X rays, Becquerel thought they might be related to fluorescence and phosphorescence. Working with a uranium salt, uranyl potassium sulfate, he found that the rays it emitted could penetrate black paper wrapped around the salt crystal and that the radiation was independent of the material's exposure to the sun, a characteristic different from the fluorescent and phosphorescent properties of the uranium salt. He also found that the radiation could ionize gases and make them conductors. This phenomenon puzzled him, as luminescence is caused by the sun's ultraviolet radiation; it stops as soon as the crystal is no longer exposed to sunlight. The radiation of uranium, however, continued even in the absence of sunlight. When Becquerel tested the

*Wilhelm Röntgen (1845-1923)*

same phenomenon on a nonluminescent compound of uranium salts, he found that the same penetrating radiation was emitted. After experimenting with a disc of pure uranium and discovering that it produced the same rays, he concluded that this radiation is unique and that it is generic to uranium and uranium compounds.

Becquerel's experiments were limited to uranium and its derivatives because he was most familiar with the salts of that element. He did not experiment with other materials, but Rutherford, following up on Becquerel's work, made efforts to understand the nature of the radiation that Becquerel had discovered. When he subjected the rays to an electromagnetic field, Rutherford found that they manifested three different characteristics:[91]

1. Rays that were barely deflected by the electromagnet and were positively charged he called alpha rays.
2. Rays that were significantly deflected by the electromagnet in the opposite direction of the alpha rays and were negatively charged he called beta rays. In 1900 Becquerel studied beta rays emitted by radium and found that the emitted particles are negatively charged and have the same value of

charge-to-mass ratio; he identified these particles as electrons.

3. Rays that were not affected by the electromagnet, and were not deflected at all, Rutherford called gamma rays. They were of an electromagnetic nature.

Within a few years it was found that beta rays are streams of electrons traveling at very high speed. In 1900 Paul Villard (1860-1934) discovered that gamma rays are more penetrating than X rays and are not influenced by magnetic fields. The nature of alpha rays, however, remained unresolved for a few years.

Becquerel's dramatic discovery of radioactivity marked the beginning of the modern approach to the study of the nucleus of an atom. It complemented the research in optical radiation leading to exploration of the atomic structure. In 1901 Becquerel published the first evidence of the radioactive transformation of one element into another.

*Marie Sklodowska Curie (1867-1934)*

In 1898 Marie Sklodowska Curie (1867-1934), a Polish physicist and mathematician, became intensely interested in Becquerel's discovery. She wrote her doctoral thesis on the evaluation of radiation characteristics of uranium. Observing that uranium ore was several times more radio-

Pierre Curie (1859-1906)

active than uranium, Madame Curie, with the assistance of her husband, Pierre Curie (1859-1906), a French physicist, found that the ore contained a new element or elements that were more active than uranium. The Curies coined the term "radioactivity" for the spontaneous disintegration of certain unstable atoms (nuclei); the emission of radiation was not fully understood at that time. In a series of experiments the Curies showed that the radioactivity of any uranium compound was directly proportional to the quantity of uranium it contained, thus demonstrating that radioactivity is an atomic phenomenon.

The Curies subjected pitchblende, a black pitchy form of the mineral uranium oxide, to systematic chemical analysis, measuring the radioactivity of the different elements obtained from pitchblende. They found two new radioactive elements, polonium and radium chloride. The radioactivity of radium was found to be more than a million times higher than that of uranium.

In 1903 Pierre Curie determined that 1 gram of radium salt released about 22,000 calories in one hour. This heat, which Curie first identified with atomic energy (spontaneous decay), is significantly less than that now produced by induced nuclear fission. Pierre Curie, fearing that radium might fall into the hands of criminals, expressed his alarm forty-two years before Hiroshima:

> *The question therefore arises whether it be to the advantage of humanity to know the secrets of nature —I am among those who believe with Nobel that humanity will obtain more good than evil from future discoveries.**

The promising career of Pierre Curie ended in 1906, when he was killed in a car accident.

By 1902 Madame Curie had isolated 0.1 gram of pure radium salt and had determined the atomic mass of radium as 225 (now confirmed at 226). In 1910, with the help of André Debierne, she finally obtained 1 gram of pure radium. The Curies also determined that the beta rays emitted by radium were negatively charged particles (electrons). In her later years Madame Curie suffered from lesions on her fingers and from anemia, the lethal effects of her prolonged exposure to nuclear radiation.

In 1904 Rutherford showed that alpha particles are helium ions. Rutherford and Frederick Saddy (1877-1956) determined that radioactive substances can be transmuted into one another. Saddy coined the word "isotopic" to describe substances that have the same atomic number z but a different mass number A; specifically they have the same number of protons but a different number of neutrons in the nucleus.† In 1912 J. J. Thomson discovered that the isotope property is not limited to radioactive elements but is shared by many other elements. With the development of a measuring device operating on the basis of charge-to-mass ratios, he confirmed that the mass of neon is roughly 20 to 22 times the mass of hydrogen.

In 1912 G. de Hevesy and F.A. Peneth,[62] two Austro-

*Pierre Curie as quoted in Robert L. Weber's *Pioneers of Science,* Institute of Physics, London, 1980, p. 18 (see reference 86).

†Atomic number Z refers to the number of protons in the nucleus and mass number A refers to the number of protons and neutrons combined.

Hungarians, invented the tracer technique by which the body's reaction to a particular chemical solution can be traced, if the element in the solution is radioactive. The body's reaction to a sodium salt, for example, can be measured if the sodium salt mixed with a radioactive salt is ingested. Measurement of the radioactivity of blood or urine samples reveals the amount of sodium absorbed by the body.

In 1919 Rutherford published a paper on nuclear disintegration, indicating that nitrogen nuclei are disintegrated by the intense force created in a close collision with alpha particles and that the proton that is liberated forms a constituent part of the nitrogen nucleus. He proposed that it would be possible to break down the nucleus structure to many lighter nuclei if a particle or similar projectiles of still greater energy were available.

Also in 1919, Otto Stern considered the possibility that hydrogen, with an atomic weight of 1.0079, might be a mixture of two isotopes. In 1931, Harold C. Urey, by using the spectrum of hydrogen, demonstrated the presence of deuterium, an isotope of hydrogen.

In 1933 Enrico Fermi accounted for the beta-decay type of radioactivity whereby a neutron changes into a proton with the creation of an electron and a neutral particle called an antineutrino.* In the most basic beta-decay process, a neutron decays into a proton and an electron.[92] The electron emitted in beta decays is not an orbital electron, nor is it an electron that was previously present within the nucleus, as the uncertainty principle eliminates the possibility that electrons exist inside the nucleus. The electron is manufactured by the nucleus out of the available energy, if the rest-energy difference between the parent and daughter† nuclei is at least $m_e c^2$, where $m_e$ is the electron mass.

In alpha decay an unstable nucleus disintegrates into a lighter

*A neutrino is a particle having negligible mass, with a spin of 1/2 and high penetration power. An antineutrino is a particle with spin 1/2 but opposite to that of a neutrino.

†Daughter nucleus is the nucleus remnant after emission of a particle from the original nucleus.

nucleus and an alpha particle (a nucleus of helium, $^4He$). Decay processes of this kind liberate energy. Because the decay products are more tightly bound than the initial nucleus, the liberated energy, which appears as the kinetic energy of the alpha particle and the daughter nucleus, may be found from the mass difference of the nuclei involved. A radioactive decay process may be part of a sequence or a series of such processes if a radioactive element of mass number A decays to another radioactive element of mass number B. The series will continue until a stable element is reached.

*Enrico Fermi (1901-1954)*

### 3.8.2 Artificial Radioactivity

In 1934 Irene Curie (1897-1956), daughter of Marie and Pierre Curie, and Frederic Joliot (1900-1958), her husband, discovered artificial radioactivity in one of the outstanding achievements of the century. They found that if aluminum foil is wrapped around radioactive polonium, the emission of positrons does not cease immediately when the active polonium is removed. The foil remains radioactive and the emission of radiation decays exponentially, as it does with ordinary radioelements. Curie and Joliot reported the same phenomenon for boron and magnesium. They found that the alpha particle impinging on aluminum produces phosphorus with a half-life of about three minutes, obtaining roughly one disintegration for every million alpha particles. The yield is so low because the aluminum nucleus repels the alpha particles by electrostatic action and prevents them from coming into contact with the target nucleus.

Through the discovery of artificial radioactivity, the dream of transmutation from one element to another became a reality. For example, an aluminum isotope $^{26}Al$ is transmuted into magnesium by emission of a positron $+\beta^+$, as follows:

$$^{26}Al \rightarrow {}^{26}Mg + \beta^+ \qquad (3.41)$$

or silver is transmuted into cadmium by emission of an electron $\beta^-$:

$$^{106}Ag \rightarrow {}^{106}Cd + \beta^- \qquad (3.42)$$

### 3.8.3 Breaking Atoms: Nuclear Fission

After the 1934 discovery of artificial radioactivity by Joliot and Curie, Fermi proposed that transmutations could be produced more effectively by bombarding nuclei with neutrons, as the neutral particles would not be repulsed by the electric field of the atom. Fermi and his co-workers Amaldi, Rosetti, Segrè, and Pontecorvo,[62] known as the School of Rome, discovered that radioactivity is enhanced when it is induced by capture of neutrons forming a radioactive isotope. The neutrons had to be slowed down first in passing through water or paraffin. Fermi and his co-workers, however, did not understand the true nature of this phenomenon.

In 1938 two German physicists, Otto Hahn (1879-1968) and his pupil Fritz Strassman (1902-1980), in one of the outstanding scientific advances of the twentieth century, discovered the true nature of the mysterious phenomenon of nuclear disintegration by working with uranium. They proposed that, when uranium is bombarded with neutrons, its nucleus is split into two equal parts and releases an energy of about 200 Mev. They believed, furthermore, that an excess of neutrons might be caused by the fission. A typical fission reaction for uranium, $^{235}U$, is as follows:

$$^{235}U + {}^{1}n \rightarrow {}^{140}Xe + {}^{94}Sr + 2\,{}^{1}n + 200 \text{ Mev} \qquad (3.43)$$

where $^{1}n$ is a neutron

$^{140}Xe$ is the element xenon

and $^{94}Sr$ is the element strontium.

This equation indicates that by impinging one neutron on an uranium nucleus two new neutrons are extracted. The excess neutrons may be used to produce still more neutrons, resulting in a divergent chain reaction. If this reaction occurs rapidly without the imposition of controls, a violent explosion results, and one has an atomic or a nuclear bomb. If the reaction occurs slowly in a controlled manner, one has an atomic power source.

The fission process reported by Hahn and Strassman was successfully duplicated and was confirmed at Columbia University by Fermi and his co-workers in 1939.[86] In the same year Edward Teller (b. 1908), Eugene Wigner (b. 1902), and Leo Szilard[14] (1898-1964), three Hungarian-born American physicists, encouraged Einstein to write his historic letter to President Franklin D. Roosevelt persuading him to sanction development of a nuclear weapon. Thus the Manhattan Project came into being, and ultimately the first atomic bomb was produced, at a cost of some $2 billion.

In December 1942 Fermi's group at the University of Chicago

achieved the first controlled self-sustaining nuclear reaction, using graphite in their reactor to moderate the energy of fission-produced neutrons. After World War II a host of new radioisotopes were investigated, among them tritium, carbon 14, uranium 233, and plutonium.

### 3.8.4 Bonding Atoms: Nuclear Fusion

In 1939 Hans Albrecht Bethe[86] (b. 1906), a German-born American physicist, proposed a theory of fusion reactions which led to the production of stellar energy by transmuting protons into helium nuclei. In the process of nuclear fusion, two light nuclei combine to form a single heavier atomic nucleus; the total mass of the product nucleus is less than the total mass of the reactants and the reduced mass is converted into energy.

Fusion takes place in all stars. From astrophysical evidence, it has been determined that the type of nucleus most abundant in the universe is that consisting of protons.[93] The transmutation of protons into helium, which is called a proton-proton chain, has the following sequence:

$$p + p \rightarrow d + e^{+} + \nu \tag{3.44}$$

$$p + d \rightarrow {}^{3}He + \gamma \tag{3.45}$$

$${}^{3}He + {}^{3}He \rightarrow {}^{4}He + p + p \tag{3.46}$$

$$p + {}^{3}He \rightarrow {}^{4}He + e^{+} + \nu \tag{3.47}$$

where p is a proton,

d is a deuteron or a nucleus consisting of one proton and one neutron,

$e^+$ is a positron (a positive electron),

$\nu$ is a neutrino (a particle without charge, having negligible mass, a spin of 1/2, and strong penetrating power),

$\gamma$ is a gamma ray,

$^3He$ is a nucleus of helium consisting of two protons plus one neutron,

and $^4He$ is an ordinary helium nucleus.

The reactions (3.44) + (3.45) + (3.47) lead to a net effect of

$$4\,p \rightarrow {}^4He + 2e^+ + 2\nu + \gamma \qquad (3.48)$$

In this process an energy of about 24.7 Mev is changed from rest mass into kinetic energy. The energy source of our sun is of this type.

Another type is the deuteron-deuteron reaction, which occurs at a much faster rate.

$$d + d \rightarrow t + p + 4.03 \text{ Mev} \qquad (3.49)$$

or

$$d + d \rightarrow {}^3He + n + 3.27 \text{ Mev} \qquad (3.50)$$

where t or $^3H$ is a symbol for triton, consisting of one proton and two neutrons.

The cross sections for these two reactions are almost equal. The reaction between deuterium and tritium has a much larger cross section and releases a much larger amount of energy per atomic mass taking part in the reaction

$$d + t \rightarrow {}^{4}He + n + 17.6 \text{ Mev} \tag{3.51}$$

The reaction between d and ${}^{3}He$ also yields a much larger amount of energy, but at a smaller cross section.

$$d + {}^{3}He \rightarrow {}^{4}He + p + 18.4 \text{ Mev} \tag{3.52}$$

Unlike the fusion process, the above process yields a product that is nonradioactive and is not a contaminant. The energy released is thus called "clean" energy.

The distance required for two nuclei to fuse is of the order of $10^{-15}$ meters. At this distance the strong nuclear force can overcome the repulsive electrostatic force stemming from the positive charges of the two nuclei. This process is particularly successful for lighter elements such as hydrogen and its derivatives, deuterium, and tritium, which have the lightest possible nuclei and therefore the lowest charge.

Controlled fusion takes place when a plasma* of deuterium (hydrogen -2) or deuterium and tritium (hydrogen -3) is heated at many million degrees centigrade. In the process the nuclei move rapidly enough to penetrate each other's electrical barriers and fuse. Deuterium is of particular interest because it constitutes about 0.015 percent of ocean water. The sun may be considered a fusion device (a thermonuclear device) because its core consists of proton fuel at a temperature of 10 to 20 million degrees centigrade.

During the early 1950s the United States developed the hydrogen bomb, or H-bomb, based on fusion reactions in a mixture of deuterium and tritium. Scientists have designed many different devices intended to make the reaction controllable under laboratory

*A plasma is a positively charged nucleus or a negative electron in a gaseous state.

conditions, but none has succeeded.

In December 1982 the first successful fusion experiment was conducted in the Tokamak Fusion Test Reactor at Princeton University. Hydrogen heated to 100,000 degrees centigrade for a fraction of a second was the gas used in the test.

### 3.8.5 Cosmic Rays

In the early 1900s, in investigating the effect of radiation from radioactive materials, researchers found that the ionization of gases could not be stopped even when the materials were removed from the vicinity of the ionization chamber. In 1903 Rutherford discovered that a radiation detector continues to respond to incoming radiation even in the apparent absence of a radioactive source and even when the detector is heavily shielded with lead. This finding suggested the existence of a mysterious radiation that, whatever its source, had great power to penetrate matter. In 1909 Gockel, measuring the intensity of this radiation at various altitudes with a balloon-mounted detector, showed that its intensity did not decrease with altitude. A year later, Victor Hess (1883-1964) and Werner Kolhoerster (1887-1945) demonstrated that, on the contrary, the radiation activity increases up to a height of 5 km.[91] The obvious conclusion was that the radiation was not of terrestrial origin. Furthermore, because the strength of the radiation did not vary between day and night, the sun could be ruled out as the culprit. It followed, then, that the mysterious rays originated in the cosmos, and they were therefore called "cosmic rays." In the 1920s Robert A. Millikan found that the water of high-altitude lakes absorbed cosmic radiation to the same degree as equal masses of low-altitude atmosphere. He therefore concluded that the earth is constantly being bombarded by highly energetic particles of cosmic rays.

In the 1930s cosmic radiation was measured at points high in the atmosphere and deep underground. A significant result of this study was the discovery, in 1937, of the mu-meson or muon. The particle, which has a mass 200 times that of an electron, an energy level of 106 Mev, and a spin of 1/2, was first identified by C. Anderson. As it was known to be capable of reaching the earth's surface, the particle evidently was not strongly interactive with atomic nuclei.

Cosmic rays originating in space, known as primary cosmic rays, are mostly protons. When primary cosmic rays enter the earth's atmosphere, their interactions with atmospheric atoms and ions give rise to secondary cosmic rays of substantially lower energy. In 1947 C. Power, C. Latters, and G. Occhelini found another new particle, whose mass was 273 times that of the electron and whose energy was about 140 Mev. Named pi-meson or pion, it had a spin of 0. A large number of other unstable particles were subsequently discovered, most of them through the use of nuclear emulsion, but some in cloud chambers developed to a high degree of sophistication.

In the 1950s, owing to the use of unmanned spacecraft and the development of solid-state detectors, scientists were able to discover the electrons trapped in the earth's magnetic field. By the early 1960s, measurements taken from the spacecraft had revealed that the so-called solar wind consists of a continuous stream of high-energy particles, including protons and electrons emitted from the sun. As they pass through the solar system at supersonic speed, some of them are trapped by the earth's magnetic field and form a belt around the earth. The belt is called Van Allen Belt in honor of James A. Van Allen, an American physicist who discovered it in 1963. The interaction of the solar wind with the earth's atmosphere produces the spectacular phenomenon of aurora borealis in the earth's polar regions.

After 1950 cosmic ray research rapidly evolved into a branch of astrophysics, joining with other branches of astronomy in probing galactic phenomena. The research has been helpful in identifying many high-energy particles, and it is still being actively pursued today.

## 3.8.6 Atomic Recoil

When an atom emits a photon the atom must recoil so as to conserve the momentum. If the atom was initially at rest its recoil momentum is then exactly equal to the photon momentum (but in the opposite direction) and its recoil kinetic energy, even for the lightest atoms, is of the order of $10^{-10}$ eV. In 1956 Rudolf Ludwig Mössbauer (b. 1929), a German physicist, discovered, by fixing atomic nuclei in the lattice of solids, that no energy may be lost in the recoil process. By locking iridium nuclei into the lattice of a crystal, Mössbauer virtually eliminated the atomic recoil and its influence on the radiated wavelength.[92]

One productive way of studying atomic systems is to perform resonance experiments, in which radiation from a collection of atoms in an excited state is incident on a collection of identical atoms in their ground state. The ground-state atoms may absorb the photons and jump to the corresponding excited state. The emitted photon energy, however, is less than the transition energy by the recoil kinetic energy. Moreover, it is less than the photon energy required for resonance, since the absorbing atom must also recoil. Because the crystalline binding energies are large, the individual atoms are held tightly to their positions in the crystal lattice and are not free to recoil; if any recoil is to occur, it must be the whole crystal that recoils. As the mass of the entire crystal is about $10^{20}$ times larger, the recoil kinetic energy becomes small and a resonant absorption may occur. The Mössbauer effect produces a gamma ray whose wavelength is invariable to one part in $10^{9}$. With it, one can detect variations in wavelength caused by gravity between the top and bottom floors of a building.

## 3.8.7 Stability of Nuclei

Certain nuclei are unstable and transform spontaneously through a decay process. When such nuclei transform into stable nuclei, the decay process stops. The transformation may be accomplished by two different decay processes, alpha decay and beta decay, which change both Z and N of a nucleus. Nuclei can, however, emit photons or nuclear gamma rays to make the transition from an excited state to the ground state without any change in either Z or N. There are many stable nuclei in nature which do not decay into a different nucleus, but no stable nuclei beyond Z = 92 have been discovered in nature.[93] The main reason is that elements beyond Z = 92 are unstable and decay into a stable nucleus with Z value lower than 92. Any element beyond the Z = 92 value must be made artificially through the neutron capture process. Theoretical evidence at this time suggests that a nucleus with Z = 114 could be stable. The instability of intermediate nuclei between Z = 92 and Z = 114, however, makes it impossible to reach a stable nucleus artificially by the process of neutron capture by lighter elements.

## 3.8.8 Dating the Past

A number of naturally occurring radioactive isotopes are not part of the decay chain of the heavy elements. As a result of nuclear reactions between air molecules and the high-energy cosmic rays, few radioactive elements are being produced continuously in the earth's atmosphere. The most notable and useful of these radioactive elements is $^{14}C$, which beta-decays with a half-life of 5,730 years.

When a living plant absorbs $CO_2$ from the atmosphere, a small fraction (about 1 in $10^{12}$) of the carbon atoms will be $^{14}C$, and the remainder will be stable $^{12}C$ (99 percent) and $^{13}C$ (1 percent). When the plant dies its intake of $^{14}C$ stops, and the $^{14}C$ decays. If it is assumed that cosmic radiation has remained at its present intensity for 20,000 to 30,000 years, and if the carbon reservoir has not changed appreciably during that period, a complete balance between the rate of disintegration of radiocarbon atoms and the rate of assimilation of new radiocarbon atoms for all material in the life cycle is soon reached. A determination of the $^{14}C$ presence in one gram of a substance, together with a knowledge of its half-life, should yield the age of that substance. In 1946, using the above reasoning, Libby[94] estimated the age of tree rings, a Bible, redwood, and other things. Carbon dating has improved since 1946, and not all of Libby's predictions were correct, though they were accurate for the time when they were made. Other methods of dating, such as potassium-argon, rubidium-strontium, and phenium-osmium, have also been widely used.

Radioactivity has been useful in many areas of medicine. In the early 1940s Ernest Lawrence and his brother, John Lawrence, from the medical school at Yale University, applied radiation to the study of biological and medical problems. They developed the clinical use of neutrons in cancer therapy. After World War II application of radiation in medicine increased significantly. Radioiodine has been effective in the diagnosis and treatment of hyperthyroidism. As noted above (see page 126), radioisotopes of many elements in the human body have been used as tracers in the study of the blood and nutrition.

## 3.9 Four Forces of Nature

By the end of the nineteenth century two forces of nature had been recognized: gravitation and electromagnetism. Gravitation was the first to be identified. Galileo subjected it to experimental study, and its precise mathematical laws were formulated by Newton. Gravitation, an attractive force that acts on all masses in the universe, is observable in the ordinary world. Its range is infinity but its strength is much less than that of the other forces.

After gravitation, electromagnetism was the second natural force to be recognized. Early experiments in electricity and magnetism were conducted by Oerestèd, Ampère, Gauss, and Faraday, who also attempted to formulate precise theories about these phenomena. Maxwell's theory of electromagnetism would eventually unify electricity and magnetism as a single force and provide a mathematically precise set of laws for it. Electromagnetism is also observable in the ordinary world and its range is also infinite. Its source is electric charges, which may be either positive or negative. Unlike gravity, electromagnetism may be either attractive or repulsive. Its frequencies range from those of radio waves to infrared, visible and ultraviolet light, X rays, and gamma rays. (For the characteristics of electromagnetic waves, see section 2.3.)

During the first thirty years of the twentieth century physicists began to wonder what forces hold atomic particles together and what forces explain the breakup of atoms which occurs in radioactivity. This line of inquiry led to the hypothesis that there are two forces in addition to gravity and electromagnetism: the nuclear strong force and the nuclear weak force. The strong force is what holds particles together in atomic nuclei. Its existence was first suggested in 1932 after the discovery of the neutron.[64] Atomic nuclei contain protons and neutrons bound to one another at distances of the order $10^{-15}$ to $10^{-14}$ meters. Experimental and theoretical arguments showed that a force of many mega electron volts (Mev) is required in order to hold these particles together. This number contrasts sharply with the very few electron volts required for binding electrons to atomic nuclei. Unlike gravity, the strong force exerts only short-range influence. Like gravity, it is independent of charge,* binding positively charged protons as well as neutral neutrons.

The nuclear weak force explains the process of radioactive beta decay first formulated by Fermi in 1933 and expanded by Hideki

*This statement is true as long as the particles involved are hadrons. Hadron is a general nomenclature for a combination of baryon, which includes protons and neutrons, and mesons.

Yukawa (1907-1981), a Japanese physicist, in 1935. This force, like the strong force, is not observable in the ordinary world and is extremely feeble. Again, the range of the force is of the order of $10^{-15}$ meters. The electromagnetic force is mediated by the emission of photons between electrons, whereas the weak interaction is mediated by the emission of massive negatively charged bosons* from neutrons, which then become protons while the bosons disintegrate into electrons and antineutrinos.

The first two forces, gravity and electromagnetism, have been familiar to man for more than 200 years. The strong and weak forces, on the other hand, were identified only recently, and consequently their role in nature is much less widely known. To some extent, of course, their unfamiliarity is fostered by the fact that strong and weak forces are not observable in everyday life.

Table 1 shows the relative strength of these four forces.

**Table 1**
**Range and Strength of Four Forces of Nature**

| Force | Range (m) | Strength |
|---|---|---|
| Gravity | Infinite | $10^{-39}$ |
| Electro-magnetism | Infinite | $\frac{1}{137}$ |
| Strong | $10^{-13}$ | 14 |
| Weak | $10^{-15}$ | $10^{-5}$ |

*A boson, an atomic particle with spin quantum number 0, 1, 2 . . ., may or may not have an electrical charge.

In 1963 Eugene Wigner, one of the many physicists confounded by the existence of four different forces in nature, voiced the opinion that nature cannot be so complex as to require for its explanation four different forces with no connections among them. This line of thinking initiated an intense effort to unify the seemingly distinct interactions.

In 1967 Steven Weinberg (b. 1933), an American physicist, proposed the unified quantum field theory, combining the electromagnetic and weak forces. He suggested that these two forces of nature can be explained by the same mechanism. Because a similar proposal was put forward independently by Abdus Salam (b. 1926), a Pakestani physicist, in 1968, the theory is known as the Weinberg-Salam or the electroweak theory. As one of the major accomplishments of physics in recent yers, it is analogous to Maxwell's equations unifying electricity and magnetism. Many experiments in recent decades have confirmed the Weinberg-Salam theory, among them the neutrino neutral current experiments[95] of 1976. Subsequent experiments[96] using beams of neutrinos and electrons have verified numerous predictions of the theory.

In an example of the weak interaction, a neutron decays into a proton, an electron, and a massless particle called an antineutrino. According to the early Fermi theory of weak interaction, the decay takes place in a single step, the neutron decaying immediately into a proton, an electron, and a neutrino. The Weinberg-Salam theory, however, describes weak interaction as a two-step process in which the neutron decays first into a proton and a negatively charged boson. In the second step, the boson decays into an electron and an antineutrino.

The system of equations describing the Weinberg-Salam theory is known as the "gauge theory," according to which the strength of weak interaction is comparable to the strength of electromagnetic interaction at high energy.* Recent experiments have supported the

*High energy here means energy of the order of 80 Mev or higher.

Weinberg-Salam unified theory.[97]

Before 1970 the Weinberg-Salam theory was not applicable to all types of elementary particles. In 1970 Sheldon Lee Glashow (b. 1932) remedied this shortcoming by extending the theory to include all elementary particles. Because of his contribution the theory unifying the electromagnetic and weak forces is now called the Glashow-Weinberg-Salam theory. Although it falls short of bringing together all four forces, it does move toward ultimate derivation of a unified theory to cover all physical forces under one simple law.

*Murray Gell-Mann (b. 1929)*

In 1964 Murray Gell-Mann (b. 1929) and G. Zweig[98] introduced the concept of quarks. In Gell-Mann's model all protons and neutrons are composed of three quarks, the building blocks of nucleus particles. In the same year Neeman,[99] an Israeli physicist, joined with Gell-Mann in proposing a classification scheme called the "Eightfold Way," which describes strongly interacting particles in terms of quarks. Subsequently, in 1973, a theory of color was proposed by H. Fritzsch, H. Leutwyler, and Gell-Mann[100] to explain the behavior of quarks. This theory, which is the foundation of quantum chromodynamics*

*Quantum chromodynamics is an application of the gauge theory for describing the interaction among quarks in the structure of protons and mesons.

(QCD),[101] has been confirmed in experiments conducted in the past ten years.[102]

In the past decade the exploration of the atomic world has flourished, reminding one of the atomic discoveries in the 1920s. Whereas in that decade the atomic distance of $10^{-10}$ meters was probed with light energy of only a few electron volts (ev), today distances of $10^{-18}$ meters and energy of $10^{-11}$ ev are being applied to further probes deep inside the nuclei of atoms to open up that complex little world.

Today scientists are striving to integrate electroweak with the strong force, thus unifying three of the forces of nature, an effort that does not seem to be beyond the realm of possibility. The next step is to bring in the unleashed force of gravity and thus to create the grand unified theory (GUT), the physicists' dream, which would combine into one unified law the four forces that drive the universe.

# 4

# Technological Explosion: The Second Half of the Twentieth Century

## 4.1 Introduction

The second half of the twentieth century is a period whose history is unfinished. More than a decade separates us from the year 2000, and much could happen between now and then, particularly in scientific advance. The nearly fifteen years remaining in the century are a long enough period for scientific revolutions to be born and even to reach maturity. For this reason it is premature to write a history of late twentieth-century science.

Still, we may be sure that our descendants, when writing the history of science for the period 1950-2000, will be struck by two significant developments: (1) an acceleration in the rate at which scientific discoveries occurred and profoundly deepened physical theory; (2) a set of differences separating this period from previous periods. Although the period from the seventeenth to the early twentieth century is marked by geniuses like da Vinci, Galileo, Newton, and Einstein, whose work covered several fields, such as optics, mechanics, and astronomy, in late twentieth-century science there are no outstanding figures working in a diversity of disciplines. The reason is to be found in the twentieth century's tendency toward specialization. In order to possess the needed skills and to be in firm command of their chosen fields of endeavor, scientists must not wander too far astray. They become deeply entrenched in narrow fields and cannot strive for the accomplishments in widespread fields which were characteristic of Renaissance scientists. As it is impossible to describe historical developments through the contributions of a wide array of specialized scientists, this chapter presents a picture of technical advances since 1950 in which light has been directly or indirectly instrumental. It is hoped that this procedure will keep the reader up to date without losing him in the labyrinths of specialized technology.

The scientific information available in the second half of the twentieth century is the richest ever known, yet it is the most difficult for the human mind to grasp. It is so minutely detailed, so complex, that one risks drowning in the learning process.

The period is best marked by the ingenious union of optics and electronics. Once quantum mechanics was applied to the exploration of microscopic behaviors of metals and semiconductors, the field of solid-state physics came into being; it led to the creation of a massive electronics industry. In the field of optics, research in television and photodetectors operating in the visible and invisible spectra, specifically infrared and ultraviolet, has expanded significantly. With the invention of the laser and its immediate application to medicine, science, industry, and manufacturing, research on its further uses exploded and is rapidly expanding.

The extensive technological advances of the past thirty years have started what might be considered a scientific revolution. The intricate devices produced in this era have had an unprecedented impact on science and medicine, among other fields, and therefore on human lives.

The discovery of the photoelectric effect, which was instrumental in opening up the atomic world, was probably the most important factor in uniting optics with electronics. By inserting a light-sensitive detector at the focal plane of a converging optical instrument, such as a television camera, an electrical signal commensurate with the radiation intensity of the observed scene is produced. This signal is transmitted to a remote location for reconversion into optical radiation and is then displayed on a television monitor. A photodetector, inserted at the focal plane of an optical telescope and extended in a linear or in a two-dimensional way (as in the human retina), eventually produces images from incoming visible as well as infrared and ultraviolet radiation. Since 1930 photodetectors have been implemented in the focal plane of practical optical instruments outside the laboratory. In the 1950s and 1960s detectors in linear and two-dimensional arrays were commonly used in infrared and ultraviolet, revealing images of the invisible world. By the late 1960s arrays of semiconductors were being incorporated into optical systems to expand the newly created field of electro-optics.

With the gradual evolution of electronic devices, photodetectors were integrated, along with electronic devices, into the focal plane of optical sensors. This technology, known today as focal-plane technology, has advanced to so high a degree that fabrication of thousands or millions of detectors, replicating the capabilities of the human retina, is feasible. In the 1970s the invention of charge-coupled devices (CCD) improved the focal-plane capability. The capacity of CCD's to store a charge at one point, transfer it, and then detect it on command at another point marks a notable advance in the functioning of focal-plane detector arrays.

Today, television, infrared, and laser are three of the most useful and rapidly expanding electro-optical systems. Of these the most influential and the most widely used is television. In addition to being the center of a cultural revolution, television is incorporated into aircraft, satellites, and space probes, helping to answer pertinent questions about space and its exploration.

A new and crucial technology accompanying electro-optical sensors is infrared. By providing a visual picture of a previously invisible world, infrared has become the most useful branch of contemporary optical research. The impressive advances of the past thirty-five years have made infrared technology one of the most exciting fields in science.

It is noteworthy that every object in the universe emits

electromagnetic radiation, though in varying degrees of intensity. These radiations may be concentrated in narrow bands or in wide spectral bands including X rays, ultraviolet, visible light, infrared, microwaves, and radio waves. Radiation detectors developed in the second half of the twentieth century convert this electromagnetic energy into a visual image of the radiation source. Large arrays of mirrors, lenses, or microwave antenna dishes, placed far apart from one another, provide high-resolution images of distant terrestrial and celestial objects.

During the 1960s and 1970s, earth resource satellites were equipped with infrared and microwave sensors* to discover the optical signatures of the various parts of the earth. This technique has been effective in locating mines, fisheries, crop diseases, and so on, as well as in ascertaining the nature of stars, planets, and moons. Infrared and microwave imaging sensors are as vital to science in the second half of the twentieth century as television was in the first half.

The 1960s were also marked by fabrication of and experimentation with another extraordinary optical device, the laser. Only twenty-five years old, today's laser has found its way into diverse areas of science and medicine. High-powered lasers pulsating in megawatts of power concentrated in a short time interval, in the range of ten femtoseconds ($10^{-14}$ sec), have recently been tested. Among other applications laser technology promises to be a valuable diagnostic tool for medicine and physics.

Another new development in optics is that of fiber optics, which began in the second half of the twentieth century. Here light is bent and reflected in a hairlike fiber to create a device that can view objects at inaccessible locations. In medicine, the application of fiber optics is vital for tracing disorders and diseases of the human body.

As electro-optical systems were integrated into streamlined computer technology, and as television systems and infrared sensors

*Infrared and microwave sensors refer to radiation sensors which sense the electromagnetic radiation at a longer wavelength than the visible region (beyond 0.7 $\mu m$).

were incorporated into space probes, general scientific knowledge skyrocketed. Once the microcomputer, in which computational capacity is stored on a single silicon chip, had been invented (1969), manufacture of microprocessors and minicomputers blossomed into one of the biggest industries of the country. In computer technology, scientists are in the process of completing the celebrated "superchip," which may contain more than 4 million transistors.

Through a variety of modern electro-optical devices we have realized the dream of knowing what exists beyond the range of the human eye. Where electro-optical research will take us cannot be foreseen, but as new technology is presented every day, it is no longer surprising to see the scientific miracles that are constantly being achieved.

In another branch of optics — photography and motion pictures — progress has been phenomenal. Photography, only 150 years old, is perhaps the most popular and the most widely marketed branch of optics today. The art of photography has extended into all fields of science, such as astronomy, biology, and physics. It captures phenomena occurring in all life forms. Photography's ability to capture time has intrigued mankind to such a degree that today film is one of the most valued forms of communication and entertainment. A flash of lightning, a volcanic eruption, the beauty and precision of a blooming flower, the impact of a drop of water falling onto a surface —all are part of a world that the camera has permanently recorded for present and future observers. When the light from an object is too faint for the naked eye to detect that object, a photographic plate can be exposed to the radiation from the object for a long period until the desired image is produced. Plates located at the focal plane of a telescope can produce the image of a faint object deep in the universe with a fidelity that the human eye cannot match.

Stunning astronomical achievements have also marked this period. Electro-optical sensors mounted and launched on space probes bring us close-up pictures of the other planets in the solar system and of their surface details, thus realizing one of man's oldest dreams. For the first time in history a telescope will be mounted on the space shuttle. It will be launched into outer space, beyond the disturbing effects of the earth's atmosphere, for probing cosmic history. The space telescope is expected to receive and deliver to earth images of objects never seen before — and perhaps find ancient light traveling through the universe carrying images of the period when galaxies were born.

## 4.2 A Building Block for a New Age: The Transistor

The invention of the transistor[103] was a direct result of the application of quantum mechanics to semiconductors.* In 1948 the development of the first transistor† was announced by John Bardeen (b.1908) and Walter Bratten[104] (b. 1902), two American physicists working in the Bell Laboratories. Also in 1948, William Shockley[105] (b. 1910), an American physicist, published his classic paper on junction diodes and transistors. The introduction of transistors immediately aroused interest in the fabrication of electronic devices and thus changed the entire landscape of electronic circuitry. The convenient size, weight, and power requirements for a transistor, along with the flexibility of combining passive and active components,‡ posed a formidable challenge to the vacuum tube.

The new computer field created an ideal market for the transistor and for its successor, the solid-state integrated circuit. Thus the transistor market expanded far beyond the conventional electronic and communication markets. In addition, the need of the

*Semiconductors are substances such as silicon and germanium, which have conductivity between metals and insulators.

†A transistor is an electronic device made of semiconductor material, usually silicon or germanium, which controls current flow in an electrical circuit.

‡Passive components refer to resistors, capacitors, and inductors. Active components refer to transistors and vacuum tubes.

military community and of space agencies to minimize the space required for and the weight of complex electronic circuits hastened the development of transistors.

The major event in the evolution of semiconductor technology was the development of the integrated circuit (IC) and its application to electrical circuits.[106] ICs consist of integration of transistors and other basic circuit elements on a piece of silicon. The integrated circuits of the early 1960s included only six to eight transistors, diodes, and resistors — enough to implement a simple logic function such as a "gate." These integrated circuits were built on a single chip with a surface area of a few square millimeters. The smallest features in the pattern defining the circuitry measured from 10 to 20 micrometers across. The chip communicated with the rest of the system through eight or ten terminals. The switching delay, which represents the time from input signal to output signal, was generally 20 to 40 nanoseconds. During the 1960s and 1970s it was the need for faster data processing and more capable computers that challenged semiconductor industries to make denser and faster chips.

During the 1960s the silicon technology provided a wide use for much faster and more powerful transistors. Small-scale transistors with a propagating delay of 1 nanosecond were introduced in the late 1960s. In the seventies, the advances of silicon technology led to larger and more complex integrated circuits, including microprocessors and semiconductor memories.

Advances in the art of semiconductor manufacturing over the past twenty years have been dramatic. The maximum available surface area on a chip has increased by more than ten times (to about 50 square millimeters) and the minimum feature size has decreased by a similar factor (to less than 1.5 micrometers).[107] High-speed chips have been built with more than 40,000 transistors and other circuit elements, organized into 5,000 separate circuit functions that switch in about 2 nanoseconds. A chip of this kind, made of bipolar transistors, may have 200 electrical contacts. Recently a single chip with more than 100,000 transistors has been produced. Optical lithography, a process whereby all the patterns that make up a circuit are ultimately transferred to the surface of the silicon chip, has had a significant effect on developing high-density chips.[108]

Today the design and fabrication of electronic devices are being retransformed by large-scale integration (LSI), a process whereby tens of thousands of transistors and their interconnections are manufactured simultaneously. Through this technology, virtually all the logic elements of a digital computer are fitted onto a chip of silicon measuring only a quarter of an inch on each side. Large-scale integrated circuits are fabricated with much smaller transistors, which can be packed more closely together than older transistors. The change in scale is measured crudely by counting the number of transistors that are fitted onto one chip. Small-scale integrated circuits have an average of 10 transistors per chip, medium-scale circuits have approximately 100, and large-scale integrations have between 10,000 and 20,000. Today the number of transistors per chip is approaching 100,000. Since the late 1970s, VHSICs (very high-speed integrated circuits)[109] have increased the availability of high-speed ICs.

The major accomplishment of microelectronics today is the fabrication of the various circuit elements and their interconnections on a single substrate in a single series of operations. This technology continues to make rapid progress.[110] Innovations and new discoveries in this field, taking place almost every day, touch virtually all aspects of human life.

## 4.3 The Beginning of Artificial Intelligence: The Computer

In 1946, at the University of Pennsylvania, J. Presper Eckert and John K. Mauchly[14] inaugurated the digital computer revolution by constructing ENIAC (electronic numerical integrator and computer). ENIAC contained 18,000 vacuum tubes, weighed 30 tons, and consumed a power of 140 KW. This monster could add two twelve-digit numbers in 100 microseconds. A few years later, in 1951, Eckert and Mauchly built another computer called UNIVAC (universal automatic computer). UNIVAC also used vacuum tubes. It was the first computer whose input/output was separated from computational circuitry, and it included both numerical and alphabetical information for the first time. ENIAC and UNIVAC typify the so-called first-generation computer, which used vacuum tubes.

As mentioned before, the development and production of semiconductor devices during the 1950s provided the foundation for fabricating faster and more powerful computers — the so-called second-generation computer. The availability of transistors, and the introduction of magnetic memory to store information, provided the needed capabilities. The speed of the computer increased from a few thousand instructions to hundreds of thousands of instructions per second. With the maturity of ICs in the late 1960s and 1970s, even faster and more powerful computers — the so-called third generation — became available.

In 1969 Hoff matched the computational capability of ENIAC with his microcomputer on a tiny silicon chip. He put a small central processing unit (CPU) on one chip, stored a program of instruction in memory on a second chip, and used a third chip to move data in and out of the CPU. The idea was practical because the technology was mature enough to integrate a large number of devices capable of performing large computations in a fraction of a millisecond. This development encouraged the fabrication of small-function computers, microprocessors, and minicomputers. The compactness of these computers lent itself to the development of desk computers.

One of the major contributions of LSI technology was the development of a key component of the computer industry, the microprocessor. Microprocessors combine the computational capacity of arithmetic logic and the central circuitry necessary to serve as a CPU. This circuitry corresponds to that part of the computation which interprets and executes instructions. The second contribution of LSI was the transfer of computer memory from the magnetic core to the semiconductor memory of a chip. The introduction of VLSICs (very large-scale integrated circuits) has led to still more compact computers. It is expected that the impact of VLSIC will be the introduction of more capable and powerful computers. They may set the stage for fourth-generation computers, used primarily for artificial intelligence.

The practical application of these devices has had a growing impact on electronic equipment such as radios, televisions, telephone computers, and satellites. The first microprocessor chip, a fourth-generation computer component about the size of three letters on this page and costing only a few dollars, nearly matched the computing power of ENIAC. The application of computers in almost every aspect of life is obvious. Computers affect communications, the stock market, flight schedules, space exploration, among others. In fact, there is no technical field today in which some kind of computer has not found its place.

## 4.4 Fabricating the Retina: Photodetectors

It is noteworthy that the photoelectric effect, an example of the interaction between light and matter, is partly responsible for the development of the theory of the quantization of energy in the atom and the consequent development of quantum mechanics. In the late 1940s, when quantum mechanics reached maturity, one of the many outcomes was its impact on the field of solid-state physics. As a result of advances in this field, a host of light-sensitive elements were explored and energy bands in various materials were recognized. Thus the quantum theory paid its tribute to its originator, light.

During the second half of the twentieth century the various behaviors of matter (particularly metals and semiconductors) under optical radiation have been intensely investigated. The energy bands in these materials and the wavelength of optical radiation corresponding to any particular energy gap for absorption or emission have been recognized (see section 3.4). The energy bands in various intrinsic semiconductor materials have been identified. When the intrinsic energy bands in the material could not create a desired optical frequency, an additional element (impurity) was added to the host material. The impurity could be selected so that the corresponding energy bands in the host material would match the desired optical frequency. The doped material ultimately obtained can be tuned to a variety of optical wavelengths. This flexibility in tuning to any desired frequency resulted in the creation of numerous photodetectors.

Although many photodetectors were developed before 1950, their main evolution has occurred during the second half of the twentieth century, when research in photodetectors and their application, especially to optical instruments, increased dramatically. During the past three decades, detector material development in almost the entire optical spectrum, from infrared to ultraviolet, has advanced rapidly. This progress may be summarized as follows:

During the 1950s individual detectors in the various optical spectra were fabricated and investigated. Improvement of individual detectors led to the production of series of detectors, either in a linear array or in a two-dimensional mosaic array, analogous to the retina of the human eye. As time passed the detector was integrated into electronic processing circuits and then inserted into the focal plane of optical instruments.

During the 1960s emphasis was placed primarily on the manufacturing and testing of linear arrays of detectors. These arrays were incorporated into practical electro-optical systems and then tested. In the latter part of the decade, the technology of integrated circuits (ICs) was combined with the technology of the photodetector to create a hybrid integration of photodetectors and processing electronics. This combination, influenced by the common base material for both detector and processing, soon created a monolithic approach to the so-called focal-plane technology.

During the 1970s the fabrication of both linear and two-dimensional mosaic arrays reached maturity. One invention, however, changed the fabric of the focal-plane detector. In 1970 a semiconductor device called the charge-coupled device (CCD) was proposed by Boyle and Smith.[111] CCD is a semiconductor material that can store the charge at one point and transfer and detect it at another point on command. Its performance includes a wide range of analog signal processing, as well as digital logic.

The already highly developed silicon technology was efficient enough for early fabrication of CCDs. The arrays for a compact and dense focal-plane sensor were a major step forward for detector technology in the latter 1970s.

In the early 1980s efforts have focused on a modular focal plane that combines a detector with processing electronics, either in a hybrid or in a monolithic format, using intrinsic as well as extrinsic material. Today this technology has been advancing at such a pace that the fabrication of two-dimensional mosaic detectors consisting of millions of detectors and similar to the human retina is not beyond the realm of possibility.

## 4.5 Bending Light: Fiber Optics

In fiber optics, also developed in this period, a glass fiber or transparent plastic uses the principle of total internal reflection.* Normally, the fibers are clustered together in a bundle to make a rod or a pipe as a transmission line for light.[112] The light enters on one side of the rods or pipes and can be detected from the other side. The light is confined as it travels along the straight rods or even bends around a gentle corner, resulting in an internal reflection. With such reflective properties along the sides, there is very little loss of light. The device is particularly useful for viewing images produced at

*The total internal reflection refers to the reflection geometry at the boundary of two optically transparent but different media when the incident light in the first medium is reflected back to the same medium.

inaccessible locations. Physicians, for example, use it to examine internal organs of the body. The integration of fiber optics and mosaic detectors has made electro-optical systems highly efficient as well as flexible.

## 4.6 Taming Light: The Laser

In 1917 Einstein published a paper that contained the foundations for laser operation and technology.[48] He analyzed the process of light emission and absorption, using an approach that was a drastic departure from that of classical physics. He treated the absorption and emission of light by applying statistical laws and probability theory. The deterministic theories of classical physics and Maxwellian electromagnetism were not considered.

Einstein considered a simple atom having only two energy states, $E_1$ and $E_2$. If $E_1$ is the lower energy state and $E_2$ is the higher, the transition from one state to the other can be realized by emission or absorption of light quanta:

$$E_2 - E_1 = h\nu \tag{4.1}$$

If the probability per unit time of the atom falling spontaneously* from the higher state to the lower state is A, and the probability per unit time of its passing from the lower state to the higher state under the influence of radiation is B,† then by requiring that the atom and the radiation be in statistical equilibrium, Einstein showed that the two rates are equal, A = B. That is, the number of atoms per second

*In spontaneous emission, the atom is initially in the upper state $E_2$ and decays to the lower state $E_1$ by emission of a photon.

†The probability B usually consists of two probabilities, $B_{12}$ and $B_{21}$. $B_{12}$ is the probability per unit time that an atom in state $E_1$ will undergo a transition to state $E_2$ by light-absorption process. $B_{21}$ is a similar probability that the atom in state $E_2$ will undergo a transition to state $E_1$ through stimulation by an incident photon and will produce a photon. This process is called stimulated emission.

going from the lower to the higher state is the same as the number going from the higher to the lower state. Through this concept Einstein obtained the transition probabilities and the emissivity of the radiation, as well as a complete derivation of Planck's radiation law. Einstein's 1917 paper, which is called "A and B of Einstein" or "Einstein's Coefficients," is now recognized as a classic in physics.

Einstein's B coefficient became the foundation for the description of stimulated emission of radiation, which is the basis of laser radiation. In the stimulated emission process, an incident photon stimulates the atom to make a transition from a higher to a lower energy state. During the process photons of the same frequency are emitted.* The stimulated emission from large numbers of atoms produces an intense light. Amplification of stimulated emission of radiation in an atomic system may be described by a combination of the Boltzmann law for energy distribution, the Einstein theory of radiation, and the selection rules in quantum mechanics.

Charles Townes[113] (b. 1915), an American physicist, was the first to build an amplifier that used excited molecules as the active element (in this case, ammonia). The amplification was in microwave and hence the amplifier was called MASER (microwave amplification of stimulated emission of radiation). The theory of light amplification through stimulated emission of radiation (LASER) was put forward by Townes and A. Schawlow (also an American physicist) in 1954. The practical laser operation was achieved in 1960 by Theodore H. Maiman,[114] who used a ruby rod as the lasing material. Since then a variety of materials have been tested as laser sources.

Laser beams revealed themselves as light sources of unprecedented intensity, collimated to a high degree. The concept that a light beam can vaporize the hardest material fulfills one of man's wildest

*The stimulated emission or radiation in an atom resembles the stimulated mechanical system resonating under the influence of a periodic mechanical force with a frequency equal to the natural frequency of the system. The stimulated emission therefore amplifies (analogous to the mechanical system) the incident photon. If this phenomenon is magnified sufficiently, a very intense and monochromatic light will emerge.

dreams. The laser has initiated a revolution in optics and has permitted the observation of new phenomena, such as those of nonlinear optics and holography. Lasers promise to be the cutting and welding tool of the future both in industry and in medicine. It has already been used to perform microsurgery, such as welding the retina of an eye to its support. A wide varity of lasers — gas lasers, liquid lasers, and solid-state lasers — cover many regions of the electromagnetic spectrum. The intense and coherent nature of the radiation has led to the increasing application of lasers in fields such as radio astronomy, microwave spectroscopy, photography, biophysics, and communications.

## 4.7 Three-Dimensional Photography: Holography

Holography was initially proposed in 1948 by Dennis Gabor[115] (1900-1979), a Hungarian-born British physicist. Working originally to improve the electron microscope, Gabor discovered that if the diffraction pattern of an object is recorded on a photographic plate, it is possible to reconstruct the image of the original object from the plate. Although Gabor's discovery had merit, it did not reach fruition until the laser was developed. Emmett N. Leith and Juris Upatniek[116] reflected laser beams from various objects, combined them with the original beam, and recorded the results on a photographic plate. The image was then reconstructed by illuminating the plate with a laser source. Thus far holography has had no significant application in industry. Its only advantage is its three-dimensional photographic capability, which remains to be exploited.

## 4.8 Remote Vision: Electro-Optical Sensors

A fundamental change in the use of optics during the second half of the twentieth century was brought about by the introduction and integration of optical telescopes with optical detectors and electronics, launching the new era of electro-optics.[34]* Through this new field the dream of knowing what exists beyond the reach of human vision has been realized. Today, electro-optics allows us to see live images transmitted from other continents and even from other planets.

The basis for electro-optical systems is an optics — a device that collects or transmits a light beam — and a detector, which converts optical radiation to an electrical signal. Early electro-optical systems incorporated a single detector at the focal plane of an optics, which collected signals. Once the signal was focused at the detectors, an electrical signal was produced, transmitted to an amplifier, and filtered for amplification and noise reduction. After filtering, the signal was usually retransmitted to a processing network for retransmission to a distant receiver. When a signal was detected by a receiver, the original image was reconstructed.

As the field of electronics and solid-state devices matured, multiple detectors in various configurations — including linear, staggered, two-dimensional arrays in ultraviolet, infrared, and visible spectra — were being developed. Of the numerous developments in electro-optical systems, the most fascinating and the most widely marketed was television.

Television is an electro-optical sensor of endless possibilities. The television system consists of two subsystems: transmission and reception. On the transmitting end, the vidicon, an electronic camera, takes a picture of the desired scene and transforms it into electrical signals. The signals are transmitted by one antenna, received by another, amplified, and fed into a video component called a picture tube. Here the signals are reconverted from electrical to optical form for viewing.

The history of television systems began in 1820 and proceeded through a series of innovations until 1920, when the components that

*An electro-optical sensor is a sensing device in which, at its input or output or both, the radiation is received or transmitted by an optical device. The optical radiation generates or is originated by an electrical signal. Television is a typical example of an electro-optical system.

make up today's television system were invented. In 1925 the list of components necessary for a TV system was completed by an American inventor, Vladimir K. Zworykin[2] (1889-1982), who conceived the iconoscope, an electronic beam-sensing camera. In recent years the development of the video tape and the video disc, which exemplify the overall operational capabilities of TV components, including high-resolution camera and color capacities, has had a significant impact on the media.

Television systems have also been used on satellites, enhancing space probes during the past twenty-five years. Remote sensory devices, consisting of a television-type camera or an electro-optical sensor operating in the infrared or ultraviolet spectrum, are mounted on a satellite or an aircraft. The infrared spectrum is used in space probes to measure temperatures and to determine composition of an alien planet by the spectral signatures of its component materials. The effective element in such probes is an electro-optical sensor, the sensing device that detects, measures, or tracks a particular spectral band, records it, and transmits it as a coded signal.

The infrared region has been particularly useful in night vision devices which sense thermal images of the scene. Thermal imaging sensors of particular use have been those operating in the mid wavelength of infrared (MWIR) (i.e., in the range of 3-5 micrometer band) and long wavelength infrared (LWIR) (i.e., in the range of 8-12 micrometer band). There is considerable absorption of infrared radiation by the atmosphere owing to atmospheric constituents such as water vapor ($H_2O$), carbon dioxide ($CO_2$), ozone ($O_3$), and methane ($CH_4$), among others. Most infrared sensors therefore operate in the atmospheric windows.

Satellites equipped with optical instruments send back information vital for making accurate weather predictions, for locating mineral deposits, for discovering concentrated fisheries in our vast oceans, and for spotting crop disease.

Electro-optics is, perhaps, one of the most complex fields of engineering. Not only does it demand simultaneous application of optics, electronics, and solid-state physics, but it also requires detailed knowledge of other fields, such as heat transfer, signal processing, data transmission, display, and atomospheric phenomenology.

During the last thirty years, the design of optical instruments has been dramatically improved by the use of computers. The variation of optical image quality as a function of the field of view and spectral bands can be evaluated through computer programs. In the past, the field of view of a telescope was a small one, as looking at celestial bodies did not require a large field of view. At present, because of a wide application of sensors in aircrafts and space probes, wide-angle fields of view and high-image quality optics are required. This necessitates the development and use of large refractive elements (lenses). Large optical elements made of glass, silicon, and germanium have been tested and used as lenses. To keep a high-image quality in optics during temperature variations, corrective elements have been included in the design. The positions of these elements are monitored by microprocessors. Larger fields of view have been achieved when a scanning mechanism is incorporated into the design. To accomplish this task, either the entire telescope was rotated on an axis or a mirror was installed inside the optical device to provide the desire scanning motion.

Single linear and two-dimensional mosaic arrays for detectors, operating in various spectral bands, have been developed, and research in this area of optics is moving forward rapidly. The sensitivity of these detectors and the reduction of their noises are improved by cooling them to cryogenic temperature. Liquid nitrogen ($77^0$ K), liquid hydrogen ($20^0$ K), and liquid helium ($4^0$ K) are used as coolants for detectors located at the focal-plane of the optical device. In space, cooling is made possible by proper baffling of optics and by taking advantage of the temperature of space ($4^0$ K). In some instances, active cooling elements have been incorporated.

Accompanying focal-plane detectors are electronic and processing circuits which are either housed with the detector at the focal plane or installed as a separate compartment. When housed with detectors, these electronic elements can be fabricated monolithically with the detectors or connected to detectors as a hybrid component. As advances are made in fabricating sensitive detectors, the monolithic approach is preferred. At present the goal is to house large numbers of detectors (possibly thousands or even millions) with their corresponding processing of electronics in the focal plane of the optics.

The data sensed by the optics and the detectors are transmitted to a receiving station after processing. For a television system, for example, the typical receiving station is the home receiver. The data thus received are processed and then presented on a display monitor.

Transmission of light through space is always altered by attenuation and scattering by the gases and turbulence in the atmosphere. Scientists have tried during the past three decades to evaluate atmospheric effects on the wide spectrum of electromagnetic radiation, including ultraviolet, infrared, and visible rays. As a result, atmospheric distortion has been corrected or accounted for to improve the performance of electro-optical systems. Because of the effects of atmospheric turbulence, light beams passing from one area of the atmosphere to another encounter two different refractive indices. The change in beam direction thus generated is referred to as "beam dancing" or "beam wobbling." These effects are especially important in laser-beam studies, where directionality is vital for successful results. At present several computer programs, notably Lotran and Hitran,[117] evaluate the effects of atmospheric absorption of various constituents as a function of optical wavelengths.

To eliminate undesired objects from the optical scene and to properly detect, process, and present the image of a desired object in the scene, a technique called spatial filtering has been devised and developed. Today's systems incorporate many spatial and electrical filtering devices, whose functions are constantly being improved.

Remote sensing optical sensors are being used in many earthbound satellites, called Earth's Resources Satellites. Through the use of these satellites the planet Earth has been photographed and mapped in detail from outer space. The cloud cover map has assisted in accurate weather predictions. The planet's marvelous cloud cover and its magnificent giant storm vortices that frequently surround our planet have produced dramatic photographs. Optical sensors incorporated in remote sensing spacecraft include sensors operating in the visible, ultraviolet, and infrared regions of the optical spectrum.

## 4.9 Trapping Time by Light: Photography and Motion Pictures

A significant outgrowth of the study of optics is photography, which has created a fascinating world of its own. By giving us photographs to look at, optics has helped to bring us into contact with the world's past and to keep us in touch with our own individual pasts. Now we can be exposed not only to distant eras, but also to distant places, peoples, feelings, and events.

Named by the Greek words for light (photos) and writing (graphos), photography began as a simple technique of recording visible images by the action of light on light-sensitive materials. It has since developed into one of man's most extensively used and most popular tools. From arts and sciences to entertainment and medicine, from still-life photography to motion pictures, the camera has penetrated areas heretofore unknown to man. Photography has reached into every facet of our lives, touching our culture, our society, by its ability to preserve the images of our times.

For centuries, scientists and artists tinkered with Leonardo da Vinci's camera obscura, a dark chamber or room with a hole in one wall through which images outside the room were projected onto the opposite wall, but it was not until the beginning of the nineteenth century that the first photographic processes took form. In 1826

Joseph-Nicéphore Niepce, a French lithographer and inventor, produced the first photograph made in a camera by using the action of light on an asphalt solution. Arrived at by a different technique, the daguerreotype was introduced by another Frenchman, artist Jacques Daguerre, who played an important role in the development of photography. In making daguerreotypes, Daguerre first treated a silver-plated sheet of copper with iodine vapors sensitive to light and then exposed it in a camera. His first plates required exposures of fifteen to twenty minutes in bright sunlight. Daguerre's process, which yielded a positive though reversed picture of milky shadows against a plain background, was the first to be commercially utilized in photography.

As more sophisticated photographic methods were discovered, however, the daguerreotype process became obsolete. Advances in photographic technology depended on the use of new chemicals, and a diversity of shapes and sizes of camera systems soon emerged. For instance, the negative-positive photographic system in use today was developed by William Talbot, a British scientist. In 1835, by soaking light-sensitive paper alternately in solutions of common salt and silver nitrate, he produced a negative from which he could make a limitless number of positives simply by putting another sheet of light-sensitive paper against the negative and exposing it to light. Although many other entrepreneurs experimented with similar techniques, Talbot and Daguerre must be credited with developing the two processes that remain the foundations of photography as we know it today.

The major difference between these two processes was that Daguerre used metal whereas Talbot used paper. Each daguerreotype was a unique creation and had excellent picture quality. Talbot's method had the advantage that his photogenic drawings could be reprinted and duplicated endlessly, though they lacked the picture quality of daguerreotypes. Nevertheless, both processes contributed to the growing fascination with photography as a way of providing a detailed record of life in its myriad facets. The criticisms of artists, who regarded photography as a shortcut to art, and of the elite, who saw little aesthetic value in a picture that could so easily be reproduced, did little to discourage its popular appeal.

By the late nineteenth century the camera was undergoing constant development. Devices became more compact and transportable, with improved lenses, more versatile shutters, and single-lens reflexes. These improvements culminated in 1875, when Eastman Kodak introduced the first roll-film camera. The company eventually expanded this concept into the comprehensive marketing scheme we know today. With the availability of small box cameras, handy rolls of negative film, and easily available processing and printing services, amateur photography rapidly became a popular pastime.

Artificial light sources were a key ancillary need in photography. An explosive product known as "flash powder" provided photographers with light until 1885, when the Austrian physicist, Ernst Mach, utilized sparks to take high-speed photographs of bullets in flight. At the Massachusetts Institute of Technology, experiments were made with devices using spark discharges and gas-discharge flash tubes. These early efforts eventually led to the electronic or speed flash so common in cameras today. In 1913 Oskar Barnack, head of the Leitz experimental department in Germany, designed a camera that became the progenitor of the 35-mm photographic technique, including comprehensive camera and accessory systems for general and scientific photography.

Color photography was first introduced, though without much success, in the mid-nineteenth century. In 1859 James Clerk Maxwell, having established the three basic photographic colors, succeeded in reproducing other colors by combining red, green, and blue light in varying proportions. The foundation for color-print processes was laid ten years later when Frenchmen Ducos du Hauren and C. Clos developed the "subtractive color synthesis," or superimposition of dye images, as a means of producing pictures in color. The process adapted what nature provided, in that the human eye is selectively sensitive to color in terms of primary stimuli of red, green, and blue light in suitable proportions. Similarly, the color-film process records the proportion of red, green, and blue colors in their mixtures with three layers of color-sensitive dye. Kodak, already profiting from a flourishing black-and-white business, was quick to seize an opportunity. Through a new subsidiary, Kodachrome, it marketed the first mass-produced color film.

Photojournalism, with subtitled on-the-scene shots of news events, blossomed in the latter part of the nineteenth century. Its most noted American practitioner was Matthew Brady, whose brilliantly conceived photographs chronicled the American Civil War. Brady's contemporary, William Henry Jackson, achieved equal eminence as a photographer of wildlife. Photography, which enabled people to reexperience history, was suddenly recognized as an art, and its practitioners were quick to increase their efforts as photographs taken in all parts of the globe won praise at exhibitions. As events of historical as well as current interest were captured and recorded in pictures, history itself became a richer and more colorful subject. As photographic technology advanced, with new processes such as duplicating, enlarging, photocopying, automations, and color and speed control, the power of the camera became evident. It could record life in a fraction of a second, reveal actions undetectable by the human eye, and reproduce images instantaneously.

The early developments in popular photography coincided with the first venture into cinema, which also attracted wide attention. The invention of moving pictures was a significant turning point in the field of photography. So simple in nature and yet so rich in content, today's movies are created simply by projecting a series of photographs at a certain speed before a viewer. The required mechanical devices include reel-to-reel film, a projector, and a mechanism to move the film through the projector at different speeds to create slow motion or high speed and even stop-action for viewing a single frame. The evolution of the first cinema machines was a gradual process, and proper applications of those machines were also discovered slowly. Their potential began to be recognized during the 1920s, but without high-speed photography there could be no motion pictures. Daguerre's first photographs required a single exposure time of fifteen minutes, whereas motion pictures require at least fifteen exposures per second. Until photographic equipment and material became fast enough and sensitive enough to make motion pictures, the development of cinema was bound to be slow.

Oddly enough, the first motion photography was the result of a bet placed in 1872. An English adventurer named Eadweard Muybridge, who followed the gold rush to California, was employed by Governor Leland Stanford to take photographs of a horse race. Stanford, an avid horse breeder and racer, had bet a rival that at one point in a galloping horse's stride all four hooves would be off the ground at the same time. Muybridge placed twelve still cameras alongside the racetrack and stretched strings across the track, each attached to the shutter of one camera. As the horse galloped down the track, his hooves broke a string and tripped a shutter, exposing one phase of his movement. Besides winning a $25,000 bet for Stanford, Muybridge became fascinated with a new kind of photography. He realized that by mounting the pictures on a revolving wheel and projecting them through light, he would resynthesize the original movement. Although this method was not the same as the one developed later, Muybridge's idea was really innovative. The true method for producing motion pictures was developed by a Frenchman, Etienne-Jules Marey, in 1882. Inspired by Muybridge's work, Marey constructed a photographic rifle that exposed twelve images in one second when the trigger was pulled. This method allowed only a few seconds of action; for longer sequences, better material and more technology were required.

During the last decade of the nineteenth century a series of inventors toyed with the idea of motion pictures, and filmstrips of "views" were actually being made and sold in France. The Lumière brothers were pioneers in the field. The business firm they established made and catalogued thousands of films of places, processes, and events. These films, however, did little to broaden the uses and applications of the camera and, when the novelty wore off, the Lumière business collapsed. By the turn of the century, shrewder businessmen and more imaginative artists, as well as the public, were evincing interest in this mechanical toy. With the development of stop-action, moving camera, close-up, and editing techniques, imagination and creativity supplanted the previous amateurish efforts. The stage was set for the emergence of motion pictures as an entertainment medium in the 1920s.

Slapstick comedies, documentaries, and unedited newsreels of world events were among the earliest ventures in the movie industry, attracting crowds of people of all ages into nickelodeons. Meanwhile, artistic and commercial development of moving pictures was progressing rapidly in Europe. Early film directors experimented with the camera in various ways, setting up outdoor chase sequences and erecting stages for elaborate moving scenery. The versatility of a moving camera was soon revealed and eventually, during the first decade of the twentieth century, new concepts were introduced. For example, stage dramas with familiar theater actors and musicians were brought to the audiences in motion picture houses. In time the cinema was recognized as a social document that could make political and philosophical statements. By the 1950s technical advances had introduced the wide screen and had led to the almost total domination of color. Creative editing, subjective camera work, and the synthesis of music, sound, and dialogue enhanced and broadened the flexibility of motion picture entertainment, which became a prosperous industry on its own.

Although new and creative ideas burgeoned, centuries-old themes were investigated as well, and cultural and traditional subjects became a central motif in motion pictures. The art of storytelling shifted to the hands of movie producers, directors, and actors. The idea of going to see a movie reigned as the entertainment event of an evening, telling the audience a story in a unique way and as an unusual and exciting experience. The modes of expressing traditional themes through motion pictures were unlimited in their ingenuity.

Photography and motion pictures have become a cultural phenomenon in the world of today. Films portray social problems, explore human psychological needs and conditions, and present sets of moral values. Film translates into cinematic terms many of the intellectual and moral questions raised in the world's great novels, plays, and philosophical essays. Once the motion picture industry was off the ground, it was exploited for many purposes, providing a medium for representing all aspects of life. The film industry grew rapidly as technological advances improved color processes and perfected sound synchronization. The development of video screens and television first competed and finally collaborated with film, making available to viewers powerful and fascinating audiovisual spectacles.

Once photography and motion pictures were installed in a high-technology category of their own during the 1960s, they were increasingly applied to scientific investigations.

Today cameras are the leading optical instruments used in science and technology. One example of their value is seen in the study of biological procreation. Various high-performance photographic techniques have been especially useful in making the wonders of reproductive biology accessible to the public. Two of the most worthwile educational uses of film today are in telescopic and microscopic viewing. Techniques to detect and project invisible or distant objects to the screen for clear viewing have dramatically transformed science and medicine.

Experiments with X rays, ultraviolet and infrared spectra, and the colors of visible light have made the use of film even more innovative. By employing the various spatial and temporal capabilities of film, man has learned a great deal about formerly invisible phenomena. Finely tuned camera equipment has made available different speeds and different magnifications for film, both of which have added another dimension to film making. The beauty and the precision of a blooming flower can be observed by viewing at high speed pictures that required hours of filming. The impact of a drop of water falling onto a surface can be observed by mangifying and slowing down a film of the event. And, by accelerating film taken by fiber optic cameras, one can actually watch the astounding, miraculous process that leads from conception to a newborn infant. Through such techniques, heretofore imperceptible natural phenomena have become visible, exhibiting the grandeur and the drama of life in all its forms. Mysteries that confounded scientists for centuries have been solved through the magic of the camera, an optical invention of the highest order.

Perhaps the most important attribute of the camera is its ability to freeze another dimension in time, thus capturing natural human emotions and evoking nostalgia and retrospection. The power of photography to call up memories brings color and richness to personal lives everywhere.

## 4.10 Nature's Optical Device: The Human Eye

Progress in optics was stimulated by man's curiosity about the mechanism of the human eye. The history of optics, as set forth in this book, began with theories of vision. It is therefore appropriate to include here a description of the present theory of vision.

For ancient man, lack of scientific knowledge based on experimental data led to speculations about the mechanism of the human eye. In ancient Greece two differing theories about the process of human vision were put forward by Democritus and Empedocles (see pages 11 to 16). For centuries, theories of vision continued to evolve, eventually forming the foundation of optical knowledge accepted today.

It was not until Leonardo da Vinci, however, that the working mechanism of the human eye was explored. Da Vinci was the first to dissect a human eye and explore its anatomy. He likened the mechanism of the human eye to his camera obscura. A major breakthrough in the visual ray theory came when he constructed a glass model of the eye and its lens so that he could look through it and ultimately confirm his theory that the optic nerve carried visual impressions. This discovery is today's basis for the understanding of the anatomy and the functioning of the human eye.

It is intriguing to know, as we now do, that we perceive the world through the decoding of complex electrical signals rather than by the direct perception of light. The light received from the outside world is collected by the eye's lens, which focuses it onto the retina. At this point the light stops, but it is transformed, through an electro-chemical process, into electrical signals that go to the brain for recognition. The perception of the outside world by the brain is accomplished by a complex decoding of these electrical signals.

The human perceptive system consists of two eyes, the neural network and pathways connecting the eyes to the brain, and the visual cortex, the part of the brain concerned with visual sensation.[118] This intricate system receives light and images and carries them to the visual center in the brain. As light enters the eye, it is refracted by the cornea, the hard front surface of the eye. The pupil, or the hole in the iris, controls the amount of light that enters the inside of the eye. The diameter of the pupil contracts and expands according to the amount of incoming light rays to the edges of the lens and thereby decreases or increases the quality of the optical image. The eye's lens, with variable focusing power, collects the incoming light rays, permitting the viewer to adjust for objects at varying distances. The rays then pass through the internal fluid of the eye, ultimately reaching the retina, where the neural imaging process begins. Information is transmitted from the retina to the brain, encoded in nervous impulses. There is, however, considerable controversy over the precise nature of the encoding, as well as over the brain's processing routines.

The retina is a complex network of interconnected cells that receive light, detect motion, register color, and, to some extent, enhance images. Much of human precognitive pattern recognition occurs or begins in the retina.

The retina consists of two distinctly different kinds of photoreceptors: rods and cones. They appear in different parts of the retina, have different sensitivities to light energy, and are connected differently to neural pathways.[119] The cones are concentrated in the center of the eye, or the fovea. The fovea encompasses 2 degrees in the center of the visual field, or about 1 degree to either side of the visual axis. Rods, virtually absent in the fovea, increase in density with distance from the fovea, reaching a maximum density about 18 to 20 degrees from the fovea. The area between 5 and 18 degrees downward from the fovea and on the horizontal meridian of the eye has no rods or cones. This area, about 3 degrees in diameter, is termed the "blind spot" and is the point at which the neural fibers from the rods and cones exit the eye. Any image falling on the blind spot cannot be seen.[120]

In each eye along the exit run about 1 million fibers wrapped tightly in a bundle, referred to as the optic nerve.[121] Each optic nerve is capable of carrying 10 billion impulses or neural firings per second.

The fundamental aspects of perception performed by the eye are depth perception, color perception, and sensitivity to motion. Although experts are in general agreement about depth, color, and motion detection, they differ widely in accounts of how man recognizes the outlines of objects, identifies these objects, senses variations in texture and density of patterns, and resolves an image into large and small details.

Depth perception seems to be little more than a comparison by the brain of the separate images it receives along the two optic nerves. There is ample evidence that depth perception or steriopsis is preattentive and that it occurs early in the visual processing system. Because the relative displacement of the images from the two eyes is seen horizontally, humans usually estimate distances on a horizontal plane with more precision than on a vertical plane or on heights.

Despite intensive study of human color vision, the mysteries of its encoding and sorting remain unsolved. Cone-shaped visual cells may be divided into at least three structural classes, suggesting the existence of a physical basis for trichromatic vision.[122] This fact is significant because human color vision is demonstrably trichromatic. Evidence from brain research further supports the idea that three distinct neural channels exist for the perception of color. There is also evidence that color processing occurs in the cortex as well as on the retina. When different colors are presented separately to the two eyes, they are sometimes perceived by the subject as a single third color. The eye is able to distinguish several hundred different spectral bands. Within the spectral bands subtle distinctions in hue can be detected.

One of the most powerful strategies used by researchers into human perception is to study the conditions under which various human systems fail. Color vision is being studied in this way. Experiments show that the same color may be subjectively perceived in different combinations of frequencies as well as in single frequencies.

Perception of motion has been the subject of experiments on certain animal species. The results have consistently led to a hierarchical model of intraneural communication. Neurons receiving the impulses that represent an image tend to change their electrical activity in a wavelike pattern as an object moves across the visual field. Such changes are evidently sensed by another, presumably "higher," group of cells. Although the existence of such hierarchies seems well proven in animals and has in fact been demonstrated in great detail in insects, the hierarchies remain hard to demonstrate in man. Recent findings indicate that individual visual cells can also sense motion; their electrical output can carry distinctive "signatures" which denote the direction of motion. Once again, the human visual processor seems to perform tasks in several places at once or globally.

## 4.11 Advances in Microscopes

New discoveries in physics in the twentieth century led to the improvement of microscopes. When de Broglie suggested, in 1923, that all matter acted as waves and waves acted as particles, the door was opened for construction of a new microscope operating on the basis of electron waves rather than light waves. By replacing light with an accelerated electron beam, microscope resolution was improved. This advance ultimately made way for the invention of the electron microscope. In 1931 accelerated electrons were first used in place of light in the design of microscopes. This first electron microscope was completed and successfully tested two years later, in 1933.

Unlike optical microscopes, electron microscopes operate through the scattering of a collimated beam of electrons by the atoms of the object under observation. Once again, unlike their optical counterparts, electron microscopes require that the electron beam occurs in a vacuum. This beam may be focused either electrostatically or electromagnetically. The image it produces is magnified and projected onto a fluorescent screen or a photographic plate.

At present, electron microscopes give reliable point-to-point resolutions to less than three angstroms (one agstrom is one ten-billionth of a meter), permitting direct observation of large molecules. Magnifications as high as 50,000 have been achieved. Electron microscopes have recently used electron beams to scan the surfaces of objects, projecting highly detailed images. Today microscopes have diverse applications; many scientific inventions have been incorporated into the concept of magnification, making microscopes invaluable, especially to medicine.

Another type of microscope is the stereoscopic binocular microscope, which is used for examining and assembling tiny electronic circuits and solid-state devices, as well as for complicated surgical operations. In addition, recently developed phase-contrast microscopes, along with specimen-staining techniques, have been applied to the study of living cells and to the identification of chemical compounds. A device called an image converter, which converts an image formed by invisible light, such as infrared or ultraviolet light, into a visible image, has been integrated into this microscope. It permits scientists to investigate the ultraviolet and infrared transmission or reflectivity of objects and to render visible details on objects which may be invisible in visible light.

Although applications of this technique are numerous, two particularly valuable fields are microscopic investigation of forgeries and detection of surface defects in silicon crystals used for transistors and other semiconductor devices.

In November 1984, at the IBM Zurich Research Laboratory, the most exquisitely precise and powerful microscope yet was invented.[123] Referred to as the Scanning Tunneling Microscope (STM), it has capabilities reaching far beyond those of any other magnifying instrument yet developed. The STM, with its extremely refined mechanisms, enables scientists to view surfaces of solid material, atom by atom, in units as minute as angstroms. Not only does it provide a topographical description of any kind of matter; it also reveals atomic composition and has led to better understanding of certain chemical interactions.

The principle of electron tunneling discovered in the early 1960s by Ivar Giaever (b. 1929), a Norwegian-born American physicist at General Electric, is based on the fact that some electrons from the surface of a test sample leaked out and formed an electron cloud around the sample. The concentration of electrons beyond the surface becomes exponentially lower as the distance from the surface increases. Because the electrons seem to be digging tunnels beyond the surface boundary, the effect is traditionally known as tunneling.

A scanning tunneling microscope based on this principle measures the current in an electrical circuit where the test sample is used as an electrode. The other electrode is a probe that has the configuration of a sharp needle. This needle is placed close enough to the test sample so that their electron clouds touch each other. The application of a voltage between the probe and the test sample causes the electrons to pass through a narrow channel on the electron clouds. The measured current, called the tunneling current, is an indication of the closeness of these electrodes. Thus the magnitude of tunnel current, related to the distance between the probe and the test sample, measures the roughness of the surface. Magnification higher than $10^7$ is achievable.

Such a tool has important implications in the microelectronics industry. The surface of the silicon chip, a key element in computer architecture, must be drawn sharply and accurately. Undoubtedly the STM will aid in the analysis and perfection of silicon chips.

## 4.12 Seeing the Past at Present: Astronomy

### 4.12.1 Introduction

Solar and lunar eclipses, meteor showers, comets, northern lights, and other colorful spectacles in the sky have captured man's imagination since the beginning of time. As we apply the science of optics to these astronomical phenomena, the cosmos becomes a huge laboratory for testing ever changing theories, developing new technologies, and inventing and improving optical devices to study the manifestations of light in the universe. Discoveries and inventions in the field of astronomy are among the most outstanding contributions of optics to the advancement of science and hence to the enrichment of human life. Man's curiosity about the universe has encouraged innovation in magnifying celestial bodies so as to advance his knowledge of rays of light and of the functions they perform. Now, in the twentieth century, the concept of light rays has expanded to embrace a wide variety of rays, ranging from radio waves to cosmic rays.

Although optics played a key role in the generation of the atomic model by examining the microscopic structures of atoms and their detailed emission and absorption spectra (see section 3.4), it has also served as a powerful tool in the identification of the nature and origin of distant galaxies and stars.

As noted earlier, the Sumerian[124] view of the universe was a simple one, holding that the earth, flat and fixed in space, was the center of the universe. Other civilizations — Chinese, Babylonian, Egyptian — subscribed to the same theory. Strengthened by the influence of Ptolemy in the second century A.D., this simplistic belief survived until the time of Copernicus, Galileo, and Kepler in the mid-sixteenth to the early seventeenth century. Ptolemy's *Almagest* dominated astronomy in the Middle Ages in both the Islamic and Christian worlds. The geocentric theory described in *Almagest* sought above all to explain the movements of the planets. By the sixteenth century, however, Copernicus had provided a much more accurate account of planetary motion by proposing that the sun is the center of the solar system. By the beginning of the seventeenth century Copernicus's theory had superseded Ptolemy's in some quarters, but it was not universally accepted. The ultimate defeat of the geocentric view awaited evidence gathered by Galileo in the early 1600s. His observations provided forceful confirmtion of the Copernican picture and laid the foundations for modern astronomy.

On March 13, 1781, William Herschel (1738-1822), a Prussian who emigrated to England, discovered a sixth planet which he named Uranus. He used a telescope with a 6.2-inch aperture and a 7-foot length. In September 1845 John C. Adams (1819-1892), a Cambridge graduate, and Urbain J. J. Leverrier (1811-1877), a French astronomer, were independently looking for an unknown object that was disturbing the path of Uranus's orbit. In September 1846 the German astronomer Johann G. Galle (1812-1910) discovered that the disturbing object was another planet, which he called Neptune. As Neptune's orbit is beyond that of Uranus, man's knowledge of the solar system instantly expanded by more than a billion miles.

In February 1930 Clyde Tombaugh (b. 1906), an American farmer turned astronomer, found still another planet, Pluto, at an average distance of 3.7 billion miles from the sun. By finding Pluto, the ninth planet, Tombaugh completed the discovery of the planets of our solar system.

*William Herschel (1738-1822)*

The twentieth century has been marked by phenomenal advances in astronomy unmatched in any previous era. Large telescopes (100" to 200" aperture), photographic plates that can be exposed to radiation from the cosmos for long periods of time, ultraviolet, infrared and X rays, microwaves and radiowaves all have contributed to opening a new world for mankind. Among the exciting discoveries resulting from these advances are external galaxies, nebulae, novae, and supernovae.

In the 1930s radio telescopes capable of forming images from radio waves emitted by celestial objects were invented. More recently, optical telescopes have been modified to operate in ultraviolet and infrared regions of the optical spectrum.

Present astronomical observations utilize telescopes whose mirrors measure 100 to 200 inches in diameter. They are installed mostly on high mountains in order to reduce the disturbing effect of the earth's atmosphere. With the aid of such telescopes, it was discovered that the gas clouds in the Milky Way are actually other galaxies, probably tens of billions in number and certainly at enormous distances from earth. Through this discovery the boundaries of the universe expanded by a quantum leap, just as the presumed limits of our solar system had expanded not long before.

### 4.12.2 The Big Bang

In the 1920s the American astronomer Edwin Hubble (1889-1958), by using the powerful new Mt. Wilson 100-inch telescope, found that the universe is receding in all directions. When a star is moving away from the observer, the absorption or emission lines of elements are displaced toward the longer wavelength of the spectrum. This displacement is called the "red shift." By studying the spectra of galaxies, Hubble discovered a doppler shift in their light which is compatible with the red shift. He also found that the speed of recession of these galaxies increases with distance, a fact known as Hubble's law.

As a result of these observations, the big bang theory of cosmic evolution emerged.[125] According to this theory, at one time all the matter in the universe was clustered together. Between 15 and 18 billion years ago, an original explosion caused the separation of the matter and thus started a continuous expansion of the universe. The cooling-down effect, after about a billion years, condensed the matter into separate galaxies and stars. The detailed timetable of the big bang theory is presented as follows:

*At first, when the temperature was over* $10^{10}$ **K**, *the matter consisted only of the fundamental particles, such as protons, neutrons, electrons, positrons, and neutrinos.* After about 100 seconds, however, the temperature had dropped to* $10^{9}$ **K**, *and the particles began to combine to form some heavier nuclei. This nucleosynthesis continued for a few minutes, during which about 25 percent of the mass of the material formed into helium. The composition of the fireball when nuclear building ceased is thought to have been mostly hydrogen, about 25 percent helium, and a trace of deuterium.*

*For the next million years the fireball was like a stellar interior — hot and opaque, with radiation passing from atom to atom. During this time, the temperature gradually dropped to about* 3000° **K**, *and the density to about 1000 atoms/*$cm^3$*. At this point the fireball became transparent. The radiation was no longer absorbed and was able to pass freely throughout the universe. After about 1000 million years the matter condensed into galaxies and stars. Now we must emphasize that the fireball must not be thought of as a localized explosion — like an exploding superstar. There were no boundaries and no site of the explosion. It was everywhere. The fireball is still existing in a sense. It has expanded greatly, but the original matter and radiation are still present and accounted for. The substance of our bodies came from material in the fireball. We were and are still in the midst of it; it is all around us.*†

*See section 3.8.

†From G. O. Abel's *Realm of the Universe*, Saunders College, 1980, p. 374 (see reference 124).

One confirmation of the big bang theory has been made recently. According to the American astronomer George Gamow[126] (1901-1968), the matter that dispersed in the big bang originated from a fireball of intense heat and pressure. The resulting radiation decreased its temperature, which eventually fell to 3° K, where it stands at present. The 1964 measurement of space temperature made by Robert Wilson (b. 1936), an American radio astronomer, and Arno Perzias (b. 1935), a German astrophysicist, confirmed the existence of this background radiation and thus supports the big bang theory. These two radio astronomers at Bell Laboratories were set to measure radio waves emitted by the halo of gas surrounding our galaxy. While calibrating their equipment, they found something more important — the faint echo of the cataclysmic fireball or big bang in which the chemical elements were created some 15 to 20 billion years ago. Scientists are in general agreement about the broad outlines of the big bang theory. A still puzzling problem is the total amount of mass in the cosmos, which will determine whether the universe will expand forever or whether the expansion will be slowed and finally reversed by gravitational forces. If there is enough mass, contraction and compression will prevail, and the universe could end in a "big crunch," which would obliterate all matter 50 to 100 billion years hence.

## 4.12.3 The Milky Way: Our Place in the Universe

### 4.12.3.1 Historical Development

Our sun and the solar system are not completely autonomous. Instead they are part of a larger set of celestial bodies known as the Milky Way Galaxy, which comprises a huge number of stars and star systems interspersed with clouds of dust and gas. Away from terrestrial light, an observer can easily see these numerous constituents of the galaxy concentrated into a luminous, irregular band of light that wends its way across the night sky. At first appearing to be a trail of silvery dust, the bank through a telescope resolves into literally billions of stars. Light from many of these stars never reaches the earth, obscured as it is by intervening clouds of dust which account for the dark patches in the glowing band. Even starlight that does reach the earth is dimmed and reddened by interstellar clouds. Still, very little of the starlight reaching earth originates outside the galaxy.

Ancient Greek astronomers made the first recorded observations of the Milky Way. Their records point out the band along which stars seem to be concentrated. In the second century B.C. Hipparchus, a Greek astronomer, made the first visual mapping of the night sky and the course followed by the concentration of stars. More than a millennium later, in 964 A.D., the Arabian astronomer Al-Sufi again mapped the stars, preserving the Milky Way seen by Hipparchus but adding one of the three external galaxies visible to the unaided eye.

After the invention of the telescope, distant stars, galaxies, and nebulae (very dense interstellar clouds) were discovered, making way for more extensive exploration and mapping of the Milky Way. From the sixteenth until the twentieth century, the mapping began with the assumption that our sun was the center of the galaxy and, during the eighteenth century, arrayed the stars of the galaxy more or less in a circle around the sun. In 1785 Herschel, the renowned English astronomer, used an astronomical telescope in mapping the locations of celestial objects. His work provided the first detailed information about the distribution of stars in the Milky Way. Herschel found that more stars were located on the galactic plane itself than in directions perpendicular to the plane and, moreover, that the number of stars in the latter category diminished very rapidly. Herschel concluded from his findings that the galaxy extended along the galactic plane about five times as far as it did along perpendicular lines.

The invention of more powerful telescopes and the discovery of a wider optical spectrum in the twentieth century dramatically increased existing knowledge of the galaxy, enabling man to conceive of it as a spiral with four arms. Our own solar system, far from being the galactic center, lies 2/3 of the distance to the galaxy's farthest reaches. It lies in the arm of Orion, which rotates around the center of the Milky Way at a speed of approximately 200-300 km/sec.

In 1917 Harlow Shapley (1885-1972), an American astronomer, mapped some of the star clusters known at the time. Shapley, supposing that star clusters form the ancient "bony frame" of the galaxy, discovered that they occur in heaviest concentration at a point in the Milky Way far distant from the sun. Nearer to the solar system they form a much lighter concentration. This phenomenon led Shapley to the recently verified assumption that the center of the galaxy is not the sun, but rather the region in which star clusters are most densely concentrated. Once again humankind found itself removed from the center of the universe to a point nearer to its periphery.

### 4.12.3.2 Formation and Composition

It is not yet known how galaxies come into being or how they evolve. According to the big bang theory, however, all galaxies were formed simultaneously at the birth of the universe. It is believed that the original detonation led to the formation of great protogalactic nebulae or clouds from which galaxies evolved, but the details of this theory remain to be elucidated.

Viewed "edge-on," the Milky Way is roughly lens-shaped, although, as noted above, a spiral structure is identifiable within it. It contains approximately 100 billion stars, plus large amounts of interstellar dust and gas. Originally the gas was mostly hydrogen and helium; other elements were formed later. At a distance of twenty to thirty thousand light-years from the sun lies the galactic nucleus in which star density is greater than at other locations and around which the galactic arms rotate. This nucleus is surrounded by a spherical region, a halo, which contains few stars and little dust or gas. Although it has ten times the mass of the halo and the galactic disc combined, the nucleus is obscured by particles of interstellar dust and gas, which absorb both visible and ultraviolet radiation.

### 4.12.3.3 Nebulae

If the stars of the Milky Way are seen as a spiral of carefully set diamonds, nebulae like the Crab and Ring nebulae are clear fiery opals within the setting. Other nebulae appear as large weaves of lamé; still others, as nothingness. These splendid objects are the sites of stellar birth and death. Some of them, furthermore, call up the last fading memory of the protogalactic cloud from which the galaxy originated.

As noted earlier, the so-called empty space between star systems is not in fact completely empty. Such regions are instead occupied by gas and dust clouds of widely varying density. The gases in these interstellar clouds are chiefly hydrogen and helium; other elements account for no more than 4 percent of the mass. Very small particles of water ice, silicates, silicon carbide, and graphite make up the interstellar dust. The two components — gas and dust — are of quite different age and origin. The gases are remnants of the great protogalaxy which, in the course of uncountable millennia, condensed into the star systems of the Milky Way. In contrast, interstellar dust particles were fused in the interiors of stars and then ejected into space. Their appearance thus postdates the condensation of the Milky Way into the stars that it now contains.

Gas and particles of interstellar dust absorb, reflect, and scatter light from nearby stars. Even in clouds of relatively low density, the scattering phenomenon therefore exerts a dimming effect on starlight. It is more dominant in the ultraviolet, violet, and blue than in the orange and red regions of star radiation. Subsequently stars appear to be dimmer and more reddish than their original spectra. At extremely heavy densities these clouds may obscure light sources located behind them. They can do so in two ways, corresponding to two types of dust nebulae. When the cloud is not located near a star, it appears as a dark empty region in space. For many centuries, in fact, these *dark nebulae* were taken to be regions perplexingly devoid of stars. The Great Rift, a dark band found in the northern sky, also known as the "hole in the heavens," divides the Milky Way into two branches. The Great Rift nebula was long regarded as an empty region in the Milky Way. Until the middle of the nineteenth century astronomers believed that through the hole they could see beyond the Milky Way.

If, on the other hand, the dust cloud is located near a star, its particles reflect starlight, creating what are called *reflection nebulae.* They are seen either as large, azure falls of light or as blue and purple gossamers strung from star to star.

The formation of at least some interstellar gas and dust clouds is explained by the mutual gravitational attraction that even very small bits of matter exert on one another. This process is important not merely for its account of interstellar cloud formation but also for its role in the generation of stars. As clouds of gas and dust increase in mass and density, they approach the point at which gravitational forces become strong enough to cause a general collapse of nearby matter toward the source of the gravitational attraction. By a process begun in this way, nebular matter condenses into a new star, which presumably happens most often in dark nebulae; bright nebulae, as we shall see, are often the grave sites of dying stars.

A second type of nebula is formed when a dense cloud of interstellar gas is ionized by a nearby star. Small, very hot stars emit large percentages of their radiation as ultraviolet wavelengths. This radiation is absorbed by interstellar hydrogen atoms yielding ionized hydrogen nuclei and free electrons. When this occurs in a relatively dense cloud of hydrogen located near a white dwarf star, for example, a process of ionization and electron "capture" is initiated. Ultraviolet radiation from the white dwarf ionizes the interstellar cloud, producing hydrogen ions and free electrons. The electrons subsequently undergo capture by hydrogen nuclei. As these electrons descend through various energy levels they emit light as discrete spectral bands. The Balmer series of hydrogen atoms discussed in section 3.4 is typical of this emission. This process results in a neutral hydrogen atom, which then repeats the sequence of ionization, electron capture, and light emission. It is by this sequence of events that *emission nebulae* yield their opalescent light.

One vehicle by which gas clouds come to be associated with stars is the supernova phenomenon. Supernovae, which are stellar explosions far more powerful than novae, leave behind clouds of gas from the dead or dying star. A supernova is signaled to earth by a sudden increase in the brightness of the star as it begins to explode. Sometimes the supernovae subsequently achieve a brightness 100 million times that of the sun.

As stars age they exhaust their supplies of fusible hydrogen, with the result that in a very large star the stellar core contracts. The accompanying rise in the temperature at the star's core initiates fusion of elements heavier than hydrogen, thereby providing the star with a new fuel source. In fact, in the ensuing sequence of contractions the star maintains itself by fusion of progressively heavier elements.

Ultimately this process results in a stellar core made up of iron nuclei, which do not permit fusion into still heavier elements. Then a final contraction of the stellar core begins, with a force so strong that it continues until all protons and electrons have been compressed into neutrons. Contraction of the core, which now consists purely of neutrons, is finally stopped by the resistance of the neutrons to further compression. The cessation of contraction sends a shock wave out through the layers of stellar gas which envelop the core. These layers are consequently blown off the star, leaving an enormous gas cloud surrounding a very small neutron star. At this point the process of ionization and capture begins, and the formation of an emission nebula is complete.

Less cataclysmic events than supernovae may also bring about the formation of emission nebulae. One example is the so-called *planetary nebulae,* which are formed when stars eject gases that become thin clouds surrounding them. Once again, the process of ionization and electron capture begins. Such nebulae are called planetary not because they are related to planets but because, at first observation, they resemble planets.

### 4.12.3.4 From Dust to Dust: The Solar System

According to the big bang theory, formation of our solar system took place more recently. The system began to evolve 4.5 billion years ago from a hot gas cloud about one light-year in dimension, when the explosion of a neighboring supernova caused the cloud to collapse. As a result, a dense disc of hot gas was formed. As the gas continued to contract, a dense core developed at its center. Along with an increase in pressure, the temperature at the center rose until it reached 20 million degrees F. At this critical temperature the fusion of hydrogen nuclei (see section 3.8.4) started a self-sustaining nuclear process. Thus the primitive sun was born. All members of the solar system are descended from this moment and the sequence of events which led to it. The protosun would ultimately bring about the formation of moons and planets and the creation of whatever life exists in the solar system.

The sun was surrounded by a rapidly rotating disc of gas. As the supernova expanded and then cooled, its debris condensed into small solid particles which became mixed with the collapsing interstellar cloud. As the sun started to shine, its light penetrated the surrounding dust. Near the sun, the heat was sufficient to evaporate ice, leaving rocklike and metal particles. The inner planets were formed from this material. At the periphery of the solar system the temperature was too low to melt the ices, which remained along with the gases. For this reason more massive bodies, the giant planets of the solar system, formed at its outer perimeter. The leftover gas settled into remote orbits at a distance of about two light-years from the sun, almost halfway to the nearest star. In this way the Oort cloud, named for Jan Oort (b. 1929), a Dutch astronomer who discovered it in 1950, was formed (see page 194 ).

The solar system is now in middle age. It is expected to last another 5 billion years. For the first 1.5 billion years, the consumption of the sun's hydrogen fuel will increase to sustain the internal temperature and pressure. This process raises the sun's radiant heat, causing the sun to expand and thus become a brighter star.[127] The increased solar radiant heat will also raise the temperature of the planets, not excepting the earth. The sun will melt the earth's polar icecaps, causing the sea level to rise and the land masses to be submerged.

In the remaining 3.5 billion years of the sun's life expectancy most of its hydrogen fuels will be consumed and the center core will start to cool. Internal pressure will subside and the center core will collapse. Subsequently, the core temperature will increase and cause the outside layers to expand enormously. The sun will then expand until it becomes a red giant. As such it will envelop and evaporate the nearest planets, including the earth.[128] After existing for a few hundred million years longer, the red giant explodes and again becomes a small and extremely dense burning star. This cycle of star to red giant to smaller star may be repeated several times. Eventually, when all its fuel is exhausted, the sun will become a cold dead body. As its life ends, it will become a black dwarf. In this process our planet, its life, the civilization, and all of us will evaporate and become a permanent part of the original dust cloud. But who knows . . . ?

## 4.12.3.5 From the Heavens: Comets

### 4.12.3.5.1 Introduction

People normally think of comets as brilliant spheres of light hurtling through the solar system, with equally brilliant ribbons of light glowing in their slipstreams. They make their fugitive appearances and then disappear, some for long periods, some forever.

This somewhat romantic conception of comets is natural enough, for, as they approach the sun, comets do begin to reflect and radiate light. They begin as well to form the spectacular tails that carry much of the comet's romance. In the dead areas of space, however, far from any star, comets would look very different. They would not glow at all, let alone brilliantly, and they would have no wake of light streaming behind them. Instead they would appear as small celestial bodies, composed of rock and dust frozen in ice formed from water and other substances.

These small celestial bodies are the long-traveling children of the solar system's first days. They survive from the period when the solar system was still a whirling nebula whose center was a protosun. Radiation from the protosun prevented nearby nebular gases from solidifying. Because the sun's radiation was weak at the rim of the nebula, the various gases froze, trapping nebular dust in the process. It was in roughly this way that comets are thought to have originated.

If this account of the origin of comets is indeed accurate, they witnessed the birth of the solar system. Their composition records the composition of the nebula from which the sun and the planets were formed. For this reason comets are to cosmology what fossils are to archaeology. But comets may help to explain more than the history of the solar system; they may also be of assistance in explaining events in earthly history. Some scientists recently have conjectured that comets have occasionally collided with the earth. Such collisions, they believe, cause huge clouds of dust and debris to fill the earth's atmosphere, obscuring the sun. Once solar radiation could not reach earth, plant and animal life would have begun to die. Events of this kind might, therefore, explain the extinction of the dinosaur and perhaps of other species.

### 4.12.3.5.2 Historical Observations

Aristotle believed that comets originated in the earth's atmosphere. His conception remained the orthodox view for nearly two millennia, until Danish astronomer Tycho Brahe (1546-1601) disputed it in 1577. Brahe believed tht comets originate and travel at distances beyond the lunar orbit.[129] His view was, of course, the correct one.

In the seventeenth century Isaac Newton and, to a lesser extent, Robert Hooke produced the first modern theories of comet activity and motion. By this time it was well known that comets form tails as they approach the sun, but explanations for this phenomenon had not risen above the level of idle speculation. Newton and Hooke, however, suggested that tail formation is caused by a repulsive force emanating from the sun. Newton, observing that comets grow brighter as they approach their perihelions,* believed that the sun's heat vaporizes material from the comet's nucleus as the comet approaches the sun. The vaporized material forms a gas cloud around the comet's nucleus, increasing the comet's apparent size and brightness. Newton and Hooke conjectured that material from this cloud of cometary gas and dust forms the tail. As the cloud forms, a force emanating from the sun pushes some of its material out into what we describe as the comet's tail. Neither Newton nor Hooke understood the origin or the nature of the repulsive force. Nevertheless, the description of comet-tail and coma† formation was essentially correct.

*The perihelion of an orbiting body is the point in its orbit which is closest to the star or planet around which it orbits. Its aphelion is the point farthest from the star or planet.

†See page 195 for the definition of coma.

On the basis of his theory of gravitation, Newton concluded that any body moving through the sun's gravitational field must follow a circular, parabolic, hyperbolic, or elliptical path. By this time Kepler had ruled out the first of these possibilities, so Newton reduced the number of possible paths of comets to three: hyperbolic, parabolic, and elliptical. Since comets whose paths are hyperbolic or parabolic do not return, it followed that comets making repeated appearances must describe elongated elliptical orbits. Again, Newton's prediction was correct.

Newton's theory of gravitation also permitted prediction of the periodicity of comets, though Newton himself made no attempt to do so. Such prediction was left instead to the English astronomer Edmund Halley (1656-1742), to whom historical records of comet sightings in 1531, 1607, and 1682 were available. It had been thought, however, that these were sightings of three different comets. Employing Newton's theory, Halley discovered that they were in fact sightings of a single comet. He supposed the comet's orbit to be an enormously elongated one, with a perihelion inside the earth's orbit and an aphelion beyond the orbit of Saturn. Halley calculated that the time from one appearance to the next was between 74 and 79 years, suggesting that the comet would return in 1758. Halley did not live to see his prediction confirmed, but on December 23, 1758, amateur astronomer George Palitzsch sighted the comet that would thereafter be known as Halley's comet. For the first time Newton's theory had successfully been applied to a specific problem and thus confirmed.[130, 131]

Halley was in fact not the first to apply Newton's ideas to the motion of comets. That honor fell to the Finnish astronomer Anders Lexell, who studied a comet, now called Lexell's comet, which was discovered by French astronomer Charles Messier in 1769. It has made only one recorded appearance. Lexell believed that the comet had, at one time, passed through the gravitational field of the planet Jupiter. In the region through which it passed the gravitational field was strong enough to disturb the comet's orbit, bringing it into a new orbit well within the confines of the solar system. Lexell's comet was observed following this orbit for some time, but it did not remain there. Lexell conjectured that the comet's orbit eventually became an elongated one with a perihelion at an enormous distance from the sun. This elongation presumably explains the failure of Lexell's comet to reappear in the inner solar system.

In the nineteenth century astronomical observations of comets yielded data critical for both nineteenth- and twentieth-century theories of comet structure, composition, and motion. An especially valuable example was the nineteenth-century observations of a comet discovered in 1818 by German astronomer Johann Encke (1791-1865). Encke's comet has an unusually short period of 3.3 years; no comet, in fact, has been observed to have a shorter one. Thus repeated observations of the comet over a relatively short time were possible; it has been observed about fifty times since its discovery. Data from these observations have played a starring role in later attempts to explain changes that occur in comet orbital periods.

Near the middle of the nineteenth century Bessel (1784-1846), a German astronomer, and Bredichien (1831-1904), a Russian astronomer, suggested that the curving of the tails of some comets might be caused by radiation pressure emanating from the sun. At the turn of the century physicists Lebedew, Nichols, and Hull, following up on this suggestion, demonstrated in a laboratory the existence of radiation pressure. Thereafter solar radiation pressure was recognized as the principal agent in comet-tail formation.

In 1910 a prominent French astronomer, Camille Flammarion, alerted the public to the presence of cyanogen gas, a deadly poison, in the tail of Halley's comet. As the earth actually passed through the tail that year, Flammarion thought the gas might poison the entire globe. His suggestion led to outbreaks of mass hysteria in several cities. In fact, the tail, consisting of gas and plasma blown off the comet by radiation from the sun, is so tenuous that its effects on the earth were miniscule.

Similar hysteria accompanied the 1973 appearance of the comet Kohoutek. This event was important to researchers because the comet was identified while still nine months from its perihelion passage, thus permitting scientists to prepare for detailed observation of Kohoutek. The nine months also gave doomsayers and psychics time to predict that the comet would collide with the earth. The predictions about the impending collision apparently were not borne out.

Until 1949 most astronomers believed that comets were "flying sandbanks," loose agglomerations of material akin to gravel. In that year a young Harvard astronomer, Fred Whipple, realized that no flying sandbank could hold together while spewing out gases for as long as comets do. Whipple put forward a different hypothesis, that comet nuclei are composed largely of ices. These are primarily water ice, but they may also be ices of carbon dioxide, ammonia, and other compounds.[132, 133] As a comet approaches a star, the ices are transformed from the frozen state into gases in a process known as sublimation.*

Shortly after Whipple had described the composition of comet nuclei, Jan Oort[134] explained why comets appear in the inner solar system. In 1950 Oort suggested that, at the outermost reaches of the solar system, a vast cloud of comets entirely surrounds the sun. What is now called the Oort cloud begins far beyond the orbit of Pluto and extends 50,000 AU,† almost reaching the nearest star. This region was once occupied by the rim of the nebula whose offspring are the sun, the moon, and the planets. It was here that solar radiation was too weak to prevent nebular gases from freezing into comet nuclei, destined forever to be made up of matter original to the solar system.

Oort speculated that stars passing near the cloud disturb its dynamic conditions, thus causing the ejection of comets from the

*Sublimation is the process whereby substances change immediately from a solid into a gaseous state.

†One AU is the distance between the earth and the sun.

cloud toward the inner solar system. Hence the origin of the comets we observe. In one version of this view not proposed by Oort, a single "death star" returns to the Oort cloud every 29 million years, sending many comets toward the inner solar system. Because the number of comets so dispatched is very large, the likelihood that one of them will hit the earth is sharply increased. If a massive comet should indeed collide with the earth, its impact might well produce a large dust cloud diffusing through the atmosphere and darkening the planet. The shrouded earth would be plunged into a deep freeze that would be lethal to most plant and animal species. As discussed earlier, some scientists have speculated that such an event was responsible for the extinction of dinosaurs 65 million years ago.

Not all scientists, however, agree with Oort's ideas. Some astronomers point out, for example, that cometary dust is very similar to interstellar dust. On this ground they argue that comets do not originate in an Oort cloud but are instead immigrants from other solar systems. Despite these naysayers, Oort's explanation enjoys wide support. Accumulating research regarding stellar perturbation has yielded a relatively complete picture of the effects stars have on comets. This picture has tended to verify Oort's explanation of the yearly influx of comets into the solar system.[135-139]

### 4.12.3.5.3 The Nature of Comets

Comets are, in a sense, spectacular optical illusions. As already noted, comet nuclei are fragile, irregular masses composed of small pieces of rock and interstellar dust frozen into ices of water, ammonia, and other compounds. They are not large masses, sometimes measuring as little as three miles across. When we see comets, however, we do not see their nuclei at all, for they are obscured by the brilliant gaseous comae which form around the nuclei as comets approach the sun, and by the comet tails that emerge from the comae.

The process by which nature perpetrates this deception has already been suggested. As comet nuclei approach the sun, their ices sublimate, changing immediately to the gaseous state. The gases take with them the dust and the rock fragments that had been trapped within the nuclei. These particles form a cloud, the comet's coma, whose diameter may measure as much as 100,000 miles. The coma both reflects sunlight, owing to the nebular dust it contains, and radiates its own light, as its gases ionize and begin to glow.

Ionized gas and dust from the comet's coma also form its tail. These materials are blown back from the head of the comet by solar wind, the "repulsive force" about which Newton and Hooke speculated three centuries ago.[140] Every second, millions of tons of gas are ejected by the sun, forming the solar wind that moves away from the sun at an average velocity of 400 km/sec. and has an average temperature of approximately a million degrees centigrade. The gas consists chiefly of protons and electrons, with the latter flowing in currents within the wind, generating a magnetic field. Gases from the comet's coma are carried by the solar wind in a direction opposite to the direction of the sun. As this happens, electrons within the solar wind ionize the cometary gases, carrying them along the solar wind's magnetic field lines and concentrating them into glowing plasma tails. Dust from the coma is also blown back by the solar wind, but it is a diffuse dust tail not concentrated by the magnetic field. Unlike plasma tails, the dust tails reflect solar radiation but do not produce their own.

Two further phenomena result from the processes responsible for comet-tail and coma formation. The first is the phenomenon of meteorites and meteoroids. Occasionally, where the earth passes through the tail of a comet, dust and rock, originally from the comet's nucleus, enter the earth's atmosphere, becoming meteoroids. As the meteoroids accelerate earthward, friction with the atmosphere causes them to burn, and they usually disintegrate completely before reaching the earth's surface. On rare occasions, however, meteoroids are massive enough to survive the combustion caused by their descent, and the comet remnants colliding with the earth are called meteorites. Some meteorites weighing as much as 60 tons have blasted out large craters when falling to earth; one of the largest is the 4,000-foot Meteor Crater in Arizona. More common events are the meteoroid showers typically observed in April (the Lyrid shower) and November (the Leonid shower).

A cataclysm of another magnitude altogether would result if the earth collided with the head of a comet. There is no evidence that such a collision has ever occurred, but speculation about the possibility has become a popular pastime of natural historians. As noted earlier, the impact of a comet with the earth's surface might generate a dust cloud sufficient to plunge the planet into a wintry night.

The second phenomenon resulting from comet activity concerns the periods of comet orbits. As gases sublimate from a comet's nucleus, a jet force against the comet is produced. Because the nuclear surface heats unevenly, the jet force is exerted unevenly. Changes in a comet's orbital periods not caused by gravity are explained by this force.[141] In the case of Halley's comet, for example, it increases the comet's period of about four days per revolution.

## 4.12.3.5.4 The Chemical Components of Comets

Water, in the form of ice, is believed to be a major constituent of comets, but scientists also expect to detect carbon dioxide and compounds such as methane and ammonia, which could cast light on the arresting suggestion that comets "seeded" earth with the chemicals necessary to create life.

Perhaps the most important technique for research regarding comet composition is spectroscopy. By analyzing the spectrum of the light comets emit, it is possible to find their constituents. Two types of emission are found in the optical spectra of comets:

a. Continuous emission from the center of the comet.

b. Discrete emissions from molecular ions, such as those of carbon (C), hydrogen (H), oxygen (O), and nitrogen (N), with the ions $CH^+$, $OH^+$, $CO^+$, $N_2^+$, and $CO_2^+$.

It is hoped that the research investigations planned for Halley's comet in early 1986 will shed light on the origin of comets and possibly on the mystery of creation.

## 4.12.4 Advances in Telescopes

At present the world's largest telescope, with a 6-meter aperture, is located on Mt. Pastukhov in the Caucasus Mountains in the Soviet Union. Before its installation, the largest was the 200-inch telescope on Mt. Palomar in California,[142] which is still in operation. There is a 100-inch telescope on Mt. Wilson in California and a 60-inch one on Haleakala on the island of Maui in Hawaii, among others.

Although many telescopes were being used before 1950, astronomical achievement since then has been remarkable. The recent close-up pictures of the planets and their surface details owe their existence to two technologies unavailable before 1950: space probes and electro-optical sensors with remote data-transmission capability.

Presently scheduled improvements in astronomical observations are a 100-inch space telescope and a 400-inch ground telescope. The space telescope, called the Hubble telescope in honor of the great astronomer, is scheduled for launching in 1986 from the space shuttle. Without the distorting effects of the earth's atmosphere, it is expected to discover objects never detected by ground telescopes. The Hubble telescope will look deep into cosmic history, perhaps even to the period when stars and galaxies were being born. It is worth noting that the naked eye can see as far into the cosmos as 600,000 light-years.* That is, the light rays that departed from objects of this distance have taken 600,000 years in order to reach the Earth. What these light rays show us therefore are events that took place 600,000 years ago. With the aid of a ground observatory we can see as far as 2 billion light-years. The new 100-inch space telescope will enable us to see as deep into our past as 12 billion light-years.

The objects yet to be discovered will eventually be scrutinized by a planned 400-inch telescope,[143] called the "Keck telescope" in honor of its donor. Construction of the Keck telescope is to begin in 1986 and to be completed by 1990. It is to be mounted on Mauna Kea, a dormant volcano 13,600 feet above sea level on the island of Hawaii. This site was selected because it is relatively free of pollution. The aperture of the Keck telescope will be twice as large as that of its predecessor on Mt. Palomar. It is the first ground telescope whose design was influenced by the existence of powerful computers. (Its design is quite different from that of the conventional telescope.) It is not possible to make a 10-meter mirror from a single slab of glass, as it would warp under its own weight. The telescope will have a mosaic of tightly packed, hexagonal mirrors, each 6 feet in diameter and 3 inches thick. In the finished assembly there will be thirty-six mirrors, aligned with the utmost accuracy, as errors of even a few fractions of a wavelength would degrade the telescope's image quality. To eliminate the detrimental effects of outside forces such as gravity, temperature change, and high-speed winds — all of which could diminish the optical quality of the telescope — each mirror segment will be mounted on an adjustable motorized platform controlled by a computer. The number of the computer adjustments will reach a maximum of 300 per second.

It is expected that, with the aid of the Keck telescope, objects that are 12 billion light-years away will be seen. Assuming 16 billion light-years to be the life of a universe, we will be able to perceive three-fourths of the distance into the center, or the birth, of the

*One light-year is 5.9 trillion ($5.9 \times 10^{12}$) miles, the distance light travels in a year.

universe. Astronomers will thus be able to suggest more precise answers to questions about the state of the universe at the time of its creation.

### 4.12.5 In the Heavens: Observations from Space

In 1957, with the Soviet launching of Sputnik, space became physically accessible. The United States and the Soviet Union competed with each other in a race to send larger and heavier probes into space. As space exploration intensified, large optical telescopes mounted on the probes left earth's atmosphere for the first time. The solar system and distant galaxies were examined without distortion or limitation by the atmosphere. For the first time in history, man saw his own planet from space and found that it is indeed curved, not flat.

The availability of television and photodetectors and the ability to transfer data through space contributed to marked advancement in the optical instruments that assist astronomy. Ground telescopes had been limited to direct observations by the human eye or to images recorded on film. Suddenly such instruments were improved by faster, more efficient, and more sensitive detectors and were mounted on probes for the journey into space. The close-range images of other planets taken by electro-optical sensors were now transformed into electrical signals for transmission to and reconstruction on earth. This superb technical achievement brought the heavenly bodies closer to man. Objects that had been fuzzy, featureless blobs at the focal plane of a ground telescope were being revealed with clear surface features and details. Space probes made a notable dent in the mystery of the cosmos.

Theoretically, a 200-inch telescope like the one at Palomar will resolve an image size of 0.025 second. But because of the blurring effect of the atmosphere, it resolves no better than 1 second. It is rare to resolve an image size as low as 0.25 second.[124] But establishing a telescope in outer space completely eliminates the blurring effect of the atmosphere.

During the past twenty-five years television systems have been installed on satellites and space probes. Remote sensory devices mounted on a satellite or on aircraft usually consist of a television-type camera or an electro-optical sensor operating in the infrared or ultraviolet spectrum. Infrared systems have been used in many space probes to measure the temperature of other planets and to deduce the material components of an alien planet from their spectral signatures.

The numerous space probes launched by the United States and the Soviet Union during the past twenty-five years have observed both the solar system and distant galaxies. Apollo, Ranger landers, and lunar orbiters mapped the moon's surface by television camera and transmitted the information to earth. Viking space probes photographed the Martian surface and successfully transmitted the information to earth. Mariner probes photographed the surfaces of Mars, Venus, and Mercury, and also observed the comet Kohoutek. The current space probe Venus has mapped the planet Venus and its cloud layer. The Soviet space probe Venera landed on Venus and transmitted photographs of its surface back to the Soviet Union. Pioneer probes have photographed both Jupiter and Saturn at close range and transmitted the data to earth. Voyager probe took excellent pictures of Jupiter and its satellites. And scientists in the Skylab have carried out extensive ultraviolet observations of the sun and other stars.

## 4.13 The Origin and Sustainment of Life: Light Triggering Living Organisms

One of the mysteries in science which has remained unsolved is that of the origin of life on earth. According to one popular theory, light was instrumental in the emergence of living organisms and in the evolution of plant and animal life on earth. It is possible, of course, that life might have generated elsewhere, in another solar system different from our own, and somehow have been transported by microorganisms to our planet. Another minority view is that life was brought to earth by comets.[144] According to this theory, comets might have contained the organic compounds needed for developing living cells. In the tiny pool formed in a comet by the melting of some of its ice over long periods of time, the theory holds, these organic compounds evolved into a living cell. When the comet hit the planet earth, the living cells colonized the earth and thus cultivated life.

Although it is impossible to prove with certainty any of the prevalent theories, the important question is how the living organisms evolved from nonliving chemical compounds. The popular theory says that inorganic matter was transformed into live bacteria by chemical reactions caused by the phenomenon of light.

According to this theory, life developed from nonliving material in the ocean. Approximately 4 billion years ago the earth's atmosphere contained little or no free oxygen, a dangerous and poisonous gas, which would have destroyed any organic compounds then being formed.[145] Presently humans and animals are protected from the harmful effects of free oxygen by complex physical and chemical mechanisms. In the primitive earth, a reduced atmosphere, that is, one with no free oxygen, was a requirement for the evolution of life. Because there was no oxidation, no ozone layer could have formed in the atmosphere. The absence of such a layer permitted the sun's ultraviolet radiation to penetrate the atmosphere and reach the earth's surface, thus providing the energy needed for the synthesis of organic compounds from molecules of water, carbon dioxide, or ammonia.

Without free oxygen in the atmosphere to destroy them, these compounds would have accumulated in the oceans until they reached the constituency of hot diluted soup, referred to as the "primeval soup." The pattern of life driven by solar energy as it is today was thus fixed for all time on our planet, and the stage was set for true biological evolution. These organic compounds are the basic substances of proteins, which make up the genetic structures and sustain the metabolic machinery for some, or all, known living organisms. Experiments have shown that spark discharges into mixtures of hydrogen, methane, ammonia, and water, exactly like those existing on the primitive earth, gave rise to aldehydes, carboxylic acids, and amino acids — the building blocks of proteins.

The living cells in the primitive ocean must have evolved from droplets whose chemistry was marked by survival characteristics. In today's language this means a metabolism favorable to the growth of a droplet. The droplet could have contained catalysts for polymerization and could have taken molecules and energy from its surroundings to become a still larger droplet. Within the larger droplet the survivability of smaller droplets, called daughter droplets, becomes even more favorable. If the daughter droplets also contain all the catalysts the parent droplets needed for their survival, the process of life has started.

A living cell is distinct because of two properties: (1) metabolism, and (2) reproduction. The survival of a cell depends on the absorption and utilization of molecules and energy the cell obtains from its surroundings for its maintenance. These processes guarantee the cell's survival in the short run. Its long-run survivability depends on the ability to reproduce offspring with characteristic properties similar to those of the mother cell. Laboratory experiments have shown that droplets containing protein form a membrane by binding molecules to their surface, which is the beginning of individuality. The catalyst needed to encourage reactions that favored the survival of particular droplets within the primeval soup and the copying machinery to ensure that the catalysts were not lost as the droplets were broken up and dispersed by wave action and other mechanical forces must have evolved together. If such a protein droplet with a limiting membrane were to produce more of its own substance, or to reproduce, it could rightly be called an elementary organism.

The earliest organisms were dependent on outside sources of nutrition by absorbing organic matter from the primitive sea. Later, organisms acquired the ability to make their own food. These organisms were the earliest bacteria capable of storing energy as food and releasing energy when needed by the burning of that food (respiration).

Once the first bacteria were formed on earth, an evolutionary change began to take place in the nature of the atmosphere. It was transformed from a reducing atmosphere in which one of every five molecules is oxygen. One consequence of this evolution was the formation of an ozone layer in the upper atmosphere which sharply reduced the ultraviolet radiation reaching the earth's surface.

Cyanobacteria are of particular interest because, by carrying out aerobic photosynthesis, they were largely responsible for the development of an oxygen-rich atmosphere. Photosynthesis is a process that in its overall effect is the reverse of respiration. During photosynthesis the energy of sunlight is employed to make carbohydrates from water and carbon dioxide, and molecular oxygen is released as a by-product. Photosynthesis is the process that maintains life on earth; almost all energy flow on earth has its basis in photosynthesis. Without it, the basic food supply would not be replenished and the earth would lose its supply of oxygen.[146]

The evolution of photosynthesis to its present complexity enabled living organisms to become the primary producers of energy-rich molecules instead of mere consumers of those provided by prebiological processes. The first organisms capable of photosynthesis probably made use of hydrogen sulfide ($H_2S$) as a source of hydrogen atoms for converting carbon dioxide ($CO_2$) into glucose. Later blue-green algae, the predecessors of cyanobacteria, and higher plants mastered the simple photosynthesis technique of obtaining hydrogen from water, a more elaborate two-step process that released oxygen and transformed the atmosphere into the one in which all subsequent life developed.

# Epilogue

Throughout the ages man has devised a multitude of optical inventions, and with them has discovered new worlds. Through the telescope, the microscope, television, and the infrared systenm, optics has solved mysteries of the cosmos as well as of the atomic world and the world of microorganisms. It has drastically molded and changed cultural traditions and religious beliefs and dogmas. Through the microscope, assisted by X rays, lasers, and infrared technologies, optics has boosted preventive medicine and has markedly changed medical thinking and medical procedures. Through photography, motion pictures, and television, optical innovation has profoundly affected the social and cultural modes and habits of everyday life. An astonishing wealth of information is continually transmitted to the public through optical devices. These observed results of the application of optics have established the key role of this field, not only in the procedures of science and medicine, but in the entire makeup of society.

It is axiomtic that scientific achievements are frequently unpredictable, that discovery and innovation come about unexpectedly. Although we cannot see where science will take us by the turn of the century and beyond, we do know that the momentum of scientific progress is ever increasing, in optics as well as in other branches of science.

The complex atomic theory developed in the first half of the twentieth century, with all its reverberations, has led to what may be the most extensive transition in the history of science. Once the atomic mystery was unlocked in the 1920s, the consequent microscopic phenomena in the form of electronic components changed the face of technology. Today we depend on microcomputers, microprocessors, superchips, robotic devices, television, lasers, space probes, and countless other optical and electrical inventions that are restructuring our civilization. These new technologies have given us new communication systems and have broadened the range of educational facilities, making available a wealth of information that will benefit future generations as well as our own. Outside the field of physics, science has engaged, for example, in the description of DNA and RNA molecules and of their role in living organisms. In so doing science is even attempting to explain the mechanism of human heredity. Nor is natural reproduction the only sort of replication which holds its interest. At various centers of research studying artificial intelligence, computer science is attempting to create computer programs and robots which mimic human mental powers.

We can predict that scientific efforts in the immediate future will have three main objectives. The first of these is to obtain a grand unified theory (GUT). It is the physicists' dream to identify the ultimate building blocks of matter and to unify the four forces of nature under a single law. The second objective is to produce technology that makes the desired information instantly available, bringing it to business, education, government, individuals, or institutions. The third objective is to marry the physical forces of machines to the thinking power of computers to produce the technology of robotic systems. The offspring of this marriage will be intelligent machines that can match human abilities.

The industrial revolution of the eighteenth century altered familial and educational structures of society by separating members of the family, sending the men to factories and the children to school. Today, in similar fashion, the technological and scientific revolution is restructuring various aspects of daily life. As individualism is being encouraged, families are more divided than ever before. The educational system is drawing men away from factories and sending them into the mainstream of contemporary technological life. To some extent, the role of parents and teachers in the education of children has been diminished, as some of their authority is transferred to video monitors. This unprecedented change in educational procedures may yet prove to be the most influential on the character of social structure

— even more influential than the eighteenth-century revolution.

Because of the existing informational system, social formats are also undergoing constant transformation. This radical societal metamorphosis must be understood and accepted. With healthy sociological adjustments and proper scientific education, the future promises to be even more fascinating than the past, as scientific advances continue to change the character of life.

The science of optics has played an essential role in bringing about this revolutionary change. It has solved some of the mysteries of the universe and has assisted in the advancement of other sciences and technologies. By observing its evolution, one can anticipate further innovation and new discoveries in the future. Light has always been the illuminating factor in the realm of human existence. In a sense, it has marked out the course that civilization will take. It is a noiseless signal coming from the heavens. It is the serene photodrama in the heavens which initially spurred man to look outside himself. Could it be that light and its elucidating properties might unlock the mystery of the universe and of human existence? The heavens, a vast arena of the unknown, have tantalized man throughout history, inspiring him to seek answers to puzzling questions about the universe. It is not entirely coincidental that the word "enlightenment" contains the word "light." Whenever man is free for a moment from the activities of everyday life, he withdraws into himself, there to dream of the origin of the universe and lose himself in this ageless puzzle.

These dreamings have been profound, not idle, for they have brought simplicity to what at first seemed complex and unity to what seemed diverse. In the first stanza of T.S. Eliot's *La Figlia Che Piange,* the man who speaks in the poem asks a woman he has lost to

*Clasp your flowers to you with a pained surprise,*
*—Fling them to the ground and turn*
*With a fugitive resentment in your eyes;*
*But weave, weave the sunlight in your hair.*

The experience of seeing sunlight on a lover's hair is familiar. Perhaps equally familiar is the range of feelings one has while watching film footage of nuclear detonations. As guardians of present and future generations and as keepers of the planet, we are worried by these events. Nevertheless, in nuclear explosions humans are releasing the force that powers stars; humans are making starlight.

At first glance, sunlight on a lover's hair and fusion reaction do not belong together. There seems to be no reason to relate collapsing nebulae to impressionist pastels, or supernovae to dramatic Pacific sunsets. These manifestations are simply too different from each other and our reactions to them are too varied. Yet joining them in our thoughts has some justification, for a unifying similarity underlies them all. The lover's hair is matter that interacts with sunlight, which hardly explains why Eliot's protagonist wants his ex-lover to weave the sunlight in her hair, but it does explain how she does it. Impressionistic pastels portray similar interactions between light rays and matter, as the canvases selectively reflect and absorb museum light. And the most beautiful winter sunsets over the Pacific are produced by the familiar mechanism of reflection and refraction.

Reflection and refraction, of course, are very different from the fusion reactions that power stars and nuclear devices. But the latter are also interactions of rays and matter, and in the end it is interactions of this kind which explain all natural phenomena. Although the matter involved in these interactions, and the rays interacting with it, are not always the same, all explanations of physical phenomena are ultimately made in terms of the fundamental distinction.

The foregoing does not suggest that physics in any way sheds light on the feelings behind Eliot's poem or on the romance associated with colorful sunsets. It proposes, rather, that, in spite of their apparent diversity, physical events can be brought together under a very few ideas, that the physical world is not only ordered but, in a way, simple. The simplicity of the physical world is marked by both elegance and beauty. It is not the beauty we find in lovers, paintings, and nature, but it is nonetheless almost as exciting. Our dreams about light, furthermore, have brought us to this aesthetic experience, for it was thinking about light that ultimately gave us our explanatory insight.

Light, which called forth life on the planet and has helped man to solve so many mysteries, is itself a mystery. The question persists: "What is the nature of light?" We close with the poetic words of one of the great geniuses of all time, Leonardo da Vinci:

> *Look at light and consider its beauty. Close and open your eye and then look at it. What you see of it, was not before; and what there was of it, is no more.*

# Bibliography

1. Hecht, E., Zajac, A., "Optics," Addison-Wesley Publishing Co., Reading, Massachusetts, 1979.

2. Grun, B., "The Timetables of History," a Touchstone Book, New York, 1982.

3. Wertenbaker, Lael, and the editors of U.S. News Books, "Eye, Window to the World," U.S. News Books, Washington, D.C., 1984.

4. Born, M., Wolf, E., "Principles of Optics," Pergamon Press, New York, 1959.

5. Boorstin, D.J., "The Discoverers," Random House, New York, 1983.

6. Plato, "The Collected Dialogues," edited by Edith Hamilton and Huntington Cairns, Ballingen Series, LXXI, Princeton University Press, 1961.

7. Pannekoek, A., "A History of Astronomy," George Allen and Unwin, London, 1969.

8. Alkindi, "De aspectibus," Alkindi, "Tideus and Pseudo-Euklid; drei optische werke," edited by Axel Bjornbo, Teubner, Berlin, 1912.

9. Saleh Bashara Omar, "Ibn al-Haythams, Optics," Bibliotheca Islamica, Chicago, 1977.

10. Bacon, R., "Opus Maius," Part V.

11. "Leonardo da Vinci," an Artabras Book, Reynal and Co., New York.

12. Galilei, Galileo, "Dialogues Concerning Two New Sciences," Macmillan, Reprint, Dover, New York, 1914.

13. Lipson, S.C., Lipson, H., "Optical Physics," Cambridge at the University Press, 1969.

14. "Encyclopedia Britannica," a General Reference, 1985.

15. Descartes, R., "Dioptrique, Météores; Principa Philosophiae," Amsterdam, 1644.

16. de Fermat, P., "Oeuvres de Fermat" (2), page 354, Paris, 1891.

17. Grimaldi, F.M., "Physico-Mathesis de Lumine, Coloribus, et Iride," Bologna, 1665.

18. Shaw, P., "The Philosophical Works of Robert Boyle," Second Edition, Volume II, page 70, London, 1738.

19. Hooke, R., "Micrographia," page 47, 1665.

20. Newton, I., "Optiks," 1704, Reprint, Dover, New York, 1952.

21. Newton, I., "Phil. Trans." (80), page 3075, 1672.

22. Huygens, C., "Traité de la Lumière," Leyden, 1690.

23. Young, T., "Phil. Trans.," Volume 12, page 387, Royal Society, London, 1802, "Young's Works" (1), page 202.

24. Römer, O., "Mém. de l'Acad. Sci." (10), Paris, 1666-1669, 575, de Sav., 1676, page 223.

25. Bartholinus, E., "Experimenta Crystalli Islandici Disdiaclasfici," Copenhagen, 1669.

26. Fresnel, A., "Ann. Chem. et Phys." (2), 1, page 239, 1816. "Oeuvres," Volume I, pages 89, 129.

27. Fresnel, A., "Oeuvres" (2), pages 261, 479, 1816.

28. Biot, J.B., Savart, F., "Note sur le Magnetisme de la Pile de Volta," "Ann. Chim. Phys." (2), page 15.

29. Poisson, S., "Académie des Sciences de Paris," "Diffraction," 1818.

30. Cauchy, A.L., "Mém. de l'Acad. Sci." (9), page 114, (10), page 293, (18), page 153.

31. Foucault, J., "Compt. Rend. Acad. Sci." (30), page 551, Paris, 1850.

32. Fizeau, H., Breguet, L., "Compt. Rend. Acad. Sci." (30), pages 562, 771, Paris, 1850.

33. Fraunhofer, J. von, "Gilberts Ann." (56), page 264, 1817.

34. Seyrafi, K., "Electro-Optical Systems Analysis," Electro-Optical Research Co., Los Angeles, 1985.

35. Bunsen, R., Kirchhoff, G., "Untersuchungen uber das Sonnenspektrum und die Spektrum der Chemischen Elemente," Abh. Kgl. Akad. Wiss., Berlin, 1861, 1863.

36. Balmer, J.J., "Notiz uber die Spectrallinien des Wasserstoffs," Ann. Physik (3), pages 25, 80, 1885, English translation by J.B. Sykes, "A Note on the Spectral Lines of Hydrogen," in W.R. Hindmarsh, "Atomic Spectra," Pergamon Press, Oxford, 1967.

37. Lloyd, H., "Trans. Royal Irish Academy" (17), page 145, 1833.

38. Hamilton, W.R., "Trans. Royal Irish Academy" (17), page 145, 1833.

39. Cavendish, H., "Electrical Researches," edition J.C. Maxwell, pages 104-113, Cambridge University Press, 1879.

40. Ohm, G.S., "Die Galvanische Kelte, Mathematish Hearbeitet," 1827.

41. Jackson, J.D., "Classical Electrodynamics," John Wiley, New York, 1975.

42. Faraday, M., "Experimental Researches in Electricity," London, 1839.

43. Maxwell, J.C., "A Treatise on Electricity and Magnetism," Volumes I and II, Oxford, 1873.

44. Kohlrausch, R., Weber, W., "Pogg. Ann. Physik u. Chem." (2), 99, page 10, 1856.

45. Cheng, D.K., "Field and Wave Electromagnetics," Addison, Wesley Co., Menlo Park, California, 1983.

46. Hertz, H., "Electric Waves," London.

47. Michelson, A.A., "Am. J. Sci.," pages 120, 122, 1881.

48. Eisenberg, R., Resnick, R., "Quantum Physics of Atoms, Molecules, Solids, Nuclei, and Particles," John Wiley, New York, 1974.

49. French, A.P., "Special Relativity," W.W. Norton Co., New York, 1968.

50. Michelson, A.A., Morley, E.W., "Am. J. Sci.," pages 134, 333, 1887.

51. Jaseja, T.S., Javian, A., Murray, J., Townes, C.H., "Phys. Rev.," pages 133, 1221, 1964.

52. Thomson, J.J., "Phil. Mag.," Volume 5 (44), page 293, 1897, "Phil. Mag.," Volume 5 (46), page 528, 1898, "Phil. Mag.," Volume 5 (48), page 547, 1899.

53. Reif, F., "Fundamentals of Statistical and Thermal Physics," McGraw-Hill, New York, 1965.

54. Wien, W., "Wied. Ann." (52), page 132, 1894. Strutt, J.W., the III Baron Raleigh, "Phil. Mag." (49), page 539, 1900.

55. Planck, M., "Verh. d. Deutsch Phys. Ges." (2), pages 202, 237, 1900, "Ann. d. Physik" (4), 17, page 553, 1901.

56. Einstein, A., "Ann. d. Physik" (4), 17, page 132, 1905, 20, page 199, 1906.

57. Resnick, R., "Introduction to Special Relativity," John Wiley & Sons, New York, 1968.

58. Cropper, W.H., "The Quantum Physicists and an Introduction to Their Physics," Oxford University Press, New York, 1970.

59. Alvagar, T., Bailey, J.M., et al., "Phys. Letter" (12), page 260, 1964.

60. de Sitter, W., "Proc. Amsterdam Acad." (15), page 1297, 1913, (16), page 395, 1913.

61. Adler, Bazin, & Schiffer, "General Theory of Relativity," Second Edition, McGraw-Hill, New York, 1975.

62. Segrè, E., "From X Rays to Quarks," W.H. Freeman, San Francisco, 1980.

63. Alustair, I.M.R., "Quantum Mechanics," John Wiley & Sons, New York, 1981.

64. Millikan, R.A., "Vide: The Electron," University of Chicago Press, Chicago, 1917.

65. Haken, Wolf, "Atomic and quantum Physics," Springer Verlag, Berlin, 84.

66. Rutherford, E., "Phil. Mag." (6), 21, page 669, 1911.

67. Bohr, N., "Atomic Structure," "Phil. Mag." (6), 24, pages 1, 476, 857, 1913, "Line Spectra," "Mem. Acad. Roy. Sci. & Lett.," Copenhagen (D. Kgl. Danske Vidensk. Selsk. Skrifter), (8), 4, 1917 (Fasc. 1-3), Dramers, H.A., Slater, J.C., "Quantum Radiation Theory," "Phil. Mag." (6), 47, page 785, 1924.

68. Rudden, M.N., Wilson, J., "Elements of Solid State Physics," John Wiley & Sons, New York, 1980.

69. Sommerfeld, A., "Ann. des Phys." (1), page 51, 1916.

70. Gottfried, "Quantum Mechanics," Volume 1, "Fundamentals," Benjamin/Cummings Publishing Co., London, 1966.

71. Pauli, W., "Z. f. Physik" (31), page 765, 1925, English translation, "On the Connexion Between the Completion of Electron Groups in an Atom with the Complex Structure of Spectra," ter Haar, D., "The Old Quantum Theory," Pergamon Press, Oxford, 1967.

72. Matthew, P.T., "Introduction to Quantum Mechanics," Third Edition, McGraw-Hill, New York, 1974.

73. Bockhoff, F.J., "Elements of Quantum Theory," Second Edition, Addison-Wesley Publishing Co., 1976.

74. de Broglie, L., "Thèse," "Ann. de Phys." (10), 3, page 22, Paris, 1925.

75. Davisson., C.J., Germer, L.H. "Phys. Rev." (2), 30, page 705, 1927.

76. Compton, A.H., "Phys. Rev." (2), 21, page 483, 1923.

77. Schrödinger, E., "Ann. d. Physik" (4), 79, pages 361, 489, 734, 1926.

78. Davydov, A.S., "Quantum Mechanics," Second Edition, Pergamon Press, New York, 1976.

79. Dirac, P.A.M., "Proc. Roy. Soc.," A 109, page 642, London, 1925, "Hydrogen Atom," ibid, A 110, page 561, 1926, "Relativistic Quantum Electrodynamics," ibid, A 114, page 243, 1927, "The Principles of Quantum Mechanics," Clarendon Press, Oxford, 1947.

80. Morrison, M.A., Estle, T.L., Lane, N.F., "Quantum States of Atoms, Molecules, and Solids," Prentice-Hall, New York, 1976.

81. Heisenberg, W., "Physikalischen Prinzipien der Quanten Theorie," Hirzel, S., Leipzig, 1930, English translation by Carl Eckart and Frank C. Hott, "The Physical Principles of the Quantum Theory," University of Chicago, 1930, Reprint, Dover, 1949.

82. Van der Waerden, B.L., Editor, "Sources of Quantum Mechanics," Dover Publications, New York, 1967.

83. Born, M., "Bemerkungen zur Statistischen Deutung der Quantenmechanik in Werner Heisenberg und due Physik Unserer Zeit," Braunschweig, Vieweg, page 103, 1961.

84. Dirac, P.A.M., "Proc. Roy. Soc.," A 117, page 610, London, 1928.

85. Dirac, P.A.M., "The Principles of Quantum Mechanics," 4th ed., Clarendon Press, Oxford, 1967.

86. Weber, R.L., "Pioneer in Physics," Nobel Prize Winners in Physics, Institute of Physics, London, 1980.

87. Lamb, W.E., Rutherford, R.C., "Phys. Rev." (72), page 241, 1947.

88. Feynman, R.P., "The Theory of Positrons," "Phys. Rev." (2), 76, page 749, 1949, "Space-Time Approach to Quantum Electrodynamics," ibid, page 679, "Mathematical Formulation of the Quantum Theory of Electromagnetic Interaction," ibid, (80), page 440, 1950.

89. Schwinger, J., "Quantum Electrodynamics," Dover Publications, New York, 1958.

90. Bransden, B.H., Joacham, C.J., "Physics of Atoms and Molecules," Longman, New York, 1983.

91. Semat, H., "Introduction to Atomic and Nuclear Physics," 4th ed., Holt, Rinehart and Winston, New York, 1962.

92. Krane, Kenneth, "Modern Physics," John Wiley & Sons, New York, 1983.

93. Wehr, M.R., Richards, J.A., Jr., and Adair, T.W., III, "The Physics of Atom," 4th ed., Addison-Wesley Company, 1984.

94. Libby, W.F., "Radiocarbon Dating," 2nd ed., University of Chicago Press, 1955.

95. Hasert, F.J., et al., "Observation of Neutrino-Like Interaction without Muon or Electron in the Garamamelle Neutrino Experiment," Phys. Lett. 46B, 139, 1973.

96. Blietschau, J., et al., "Evidence for the Leptonic Neutral Current Reaction," Nuclear Physics, B114, 189, 1976. Faissner, H., et al., Proceedings of International Neutrino Conference, 1976, p. 223. Reines, F., et al., Proceedings of International Neutrino Conference, 1976, p. 217. Cline, D., et al., "Observation of Elastic Neutrino-Proton Scattering," Phys. Rev. Lett., 37, 253, 1976. Lee, W., et al., "Single Pion-Production in Neutrino and Antineutrino Reaction," Phys. Rev. Lett., 38, 202, 1977.

97. Prescott, C.Y., et al., "Parity Non-Conservation Inelastic Electron Scattering," Phys. Lett., 77B, 347, 1978.

98. Gell-Mann, M., "A Schematic Model of Baryons and Mesons," Phys. Lett., 8, 214, 1964. Zwig, G., Cern preprint CERN-TH-401, 1964.

99. Gell-Mann, M., and Neeman, Y., "The Eightfold Way," W.A. Benjamin, New York, 1964.

100. Fritzch, H., Gell-Mann, M., and Leutwyler, H., "Advantage of the Color Gluon Picture," Phys. Lett., 47B, 365, 1973.

101. Dodd, J.E., "The Ideas of Particle Physics," an introduction for scientists, Cambridge University Press, Cambridge, 1984.

102. Yeh, W., "From QED to QCD and from QCD to GUT," EOR-21 and 22, 1985, Electro-Optical Research, Los Angeles, California 1985.

103. Rosenberg, H.M., "The Solid State," 2nd ed., Clarendon Press, Oxford, 1983.

104. Bardeen, J., "The Transistor, A Semiconductor Triode," "Phys. Rev.," 74, page 230, 1948.

105. Shockley, W., "The Theory of p-n Junctions in Semiconductor and p-n Junction Transistors," Bell Systems Technology, 28, page 435, 1949.

106. Noyce, Robert N., "Microelectronics," "Scientific American" (September), 1977.

107. Blodgett, A., Jr., "Microelectronic Packaging," "Scientific American" (July), 1983, pages 86-96.

108. Angell, J.B., Terry, C., Barth, P.W., "Silicon Micromechanical Devices," "Scientific American" (April), 1983, pages 44-55.

109. Barna & Arpad, "VHSIC (Very High Speed Integrated Circuits)," John Wiley & Sons, New York, 1981.

110. Dohler, Bottfriend, H., "Solid State Superlattices," "Scientific American" (November), 1983.

111. Beynon, J.D.E., and Lamb, D.R., "Charge-Coupled Devices and Their Applications," McGraw-Hill, New York, 1980.

112. Elion, G.R., Elion, H.A., "Fiber Optics in Communications Systems," Marcel Dekker, New York, 1978.

113. Townes, C.H., Schawlow, A.L., "Microwave Spectroscopy," McGraw-Hill, New York, 1955.

114. Ditchburn, R.W., "Light," Academic Press, New York, 1976.

115. Gabor, D.G., "Journ. I.E.E.," Volume III, 93, page 429, 1946, ibid, 94, page 369, 1947, "Proc. Roy. Soc.," A 197, page 454, 1949, "Phil. Mag." (7), 41, page 1161, 1950, "Nature" (166), page 724, "Proc. Phys. Soc.," B 64, page 449, 1951.

116. Hecht, J., Teresi, D., "Laser, Supertool of the 1980's" Ticknor & Fields, New York, 1984.

117. McClatchey, R.A., Renn, R.W., Sleby, J.E.A., et al., "Optical Properties of the Atmosphere," Air Force Cambridge Research Lab., Cambridge, Massachusetts, 1972.

118. Alario, J., "The Electro Optics of the Human Visual Processor," Electro-Optical Research, Los Angeles, Report No. EOR-83-11, August 1983.

119. Snyder, A.W., and Menzel, R., eds., "Photoreception Optics," Springer-Verlag, Berlin, 1975.

120. Tannas, L.E., Jr., "Flat-Panel Displays and CRT's," Van Nostrand Reinhold, New York, 1985.

121. Wolff, Eugene, and Warwick, Roger, "Anatomy of the Eye and Orbit," W.B. Saunders, New York, 1976.

122. Zinn, Keith M., ed., "Ocular Fine Structure for the Clinician," Little, Brown, Boston, 1977.

123. Binnig, G., Rohrer, H., "The Scanning Tunneling Microscope," "Scientific American" (August), 1985.

124. Abel, G.O., "Realm of the Universe," Sounders College, Philadelphia, 1980.

125. Silk, Joseph, "The Big Bang: The Creation and Evolution of the Universe," W.H. Freeman, San Francisco, 1980.

126. Gamow, G.A., "Expanding Universe and the Origin of Galaxies," I Kommission has Munksgaard, Copenhagen, 1953.

127. Layzer, D., "Constructing the Universe," Scientific American Books, New York, 1984.

128. Smoluchowski, R., "The Solar System," Scientific American Library, New York, 1983.

129. Frazier, K., "Solar System," Time-Life Books, Alexandria, Virginia, 1985.

130. Brown, P.L., "Comets, Meteorites, and Man," Robert Hale & Company, 1973.

131. Moore and Mason, "The Return of Halley's Comet," W.W. Norton, New York, 1984.

132. Whipple, F.L., "A Comet Model I. The Acceleration of Comet Encke," Astrophys. J., 111, 1950.

133. Whipple, F.L., "A Comet Model II. Physical Relations for Comets and Meteors," Astrophys. J., 113, 1951.

134. Oort, J.H., "The Structure of the Cloud of Comets Surrounding the Solar System and a Hypothesis Concerning Its Origin," Bull. Astrong. Inst., Netherlands, 11:91-110, 150 (no. 408).

135. Opik, E.J., "Note on Stellar Perturbations of Nearby Parabolic Orbits," Proc. Amer. Acad. Arts. Sci., 67:171-183, 1932.

136. Yahushita, S., "Stellar Perturbations of Orbit of Long-Period Comets," Astron. Astrophy., 16:395-403, 1972.

137. Everhart, E., "The Shortage of Comets in Elliptical Orbits," in *Dynamics of the Solar System,* ed. R.L. Duncombe, Dordrecht, Holland, D. Reidel, 1978.

138. Weissman, P.R., "Dynamical History of the Oort Cloud in Comets," ed. L.L. Wilkening, University of Arizona Press, Tucson, 1982.

139. Fernandez, J.A., "Evolution of Comet Orbits under the Perturbing Influence of the Giant Planets and Nearby Stars," Icarus, 42:406-421, 1980.

140. Mendis, D.A., Hill, J.R., Haupis, H.L., Whipple, E.C., "On the Electrostatic Charging of the Cometary Nucleus," Astrophys. J., 1981.

141. Sekanian, Z., 1977, "Dust Evolution from Comets," in Space Research XVII, Proceedings of the Open Meetings of Working Groups on Physical Science, June 8-19, 1976, and Symposium on "Minor Constituents and Excited Species," Philadelphia, June 9-10, 1976, ed. M.J. Rycroft, Pergamon Press, New York.

142. Kingslaker, R., ed., "Applied Optics and Optical Engineering," Volume V, Part II, Academic Press, New York, 1969.

143. "Keck Telescope," Los Angeles Times, January 4, 1985 (also "Science Digest," May and June, 1985).

144. "Comets and the Origin of Life," Proceedings of 5th College Park Colloquium on Chemical Evolution, ed. Cyril Ponnamperuma, University of Maryland, College Park, Oct. 29-31, 1980.

145. Dickerson, R.E., "Chemical Evolution and the Origin of Life," Scientific American, September 1978.

146. Simpson, G.G., "Fossils and the History of Life," Scientific American Library, New York 1983.

# Selected Readings

Born, "Atomic Physics," 7th ed., Hafner Publishing Co., New York, 1962.

D'Arbro, "The Rise of the New Physics, Its Mathematical and Physical Theories," Volumes I and II, Dover Publishing Co., New York, 1951.

Jammer, "The Philosophy of Quantum Mechanics," John Wiley & Sons, New York, 1974.

Whittaker, "The History of the Theories of Aether and Electricity," Volume I, Thomas Nelson & Sons, New York, 1958, Volume II, Humanities Press, New York, 1973.

# Author Index

N



O



P



R



S

# Subject Index

S



T



U

# Illustration Credits

Page 11: The Bettmann Archive, New York.

Page 12: The Bettmann Archive, New York.

Page 13: The Bettmann Archive, New York.

Page 15: The Granger Collection, New York.

Page 16: The Bettmann Archive, New York.

Page 17: The Bettmann Archive, New York.

Page 18: The bettmann Archive, New York.

Page 20: The Granger Collection, New York.

Page 23: The Bettmann Archive, New York.

Page 27: The Bettmann Archive, New York.

Page 29: The Granger Collection, New York.

Page 32: The Bettmann Archive, New York.

Page 34: The Granger Collection, New York.

Page 38: The Granger Collection, New York.

Page 39: The Bettmann Archive, New York.

Page 43: The Granger Collection, New York.

Page 44: The Bettmann Archive, New York.

Page 45: The Granger Collection, New York.

Page 46: The Granger Collection, New York.

Page 48: The Bettmann Archive, New York.

Page 49: The Bettmann Archive, New York.

Page 52: The Bettmann Archive, New York.

Page 53: The Bettmann Archive, New York.

Page 54: The Bettmann Archive, New York.

Page 55: The Bettmann Archive, New York.

Page 56: The Bettmann Archive, New York.

Page 57: The Granger Collection, New York.

Page 58: The Granger Collection, New York.

Page 59: The Bettmann Archive, New York.

Page 62: The Granger Collection, New York.

Page 64: The Bettmann Archive, New York.

Page 67: The Bettmann Archive, New York.

Page 80: The Bettmann Archive, New York.

Page 87: The Bettmann Archive, New York.

Page 95: The Bettmann Archive, New York.

Page 96: The Bettmann Archive, New York.

Page 97: The Granger Collection, New York.

Page 102: The Granger Collection, New York.

Page 104: The Granger Collection, New York.

Page 108: The Bettmann Archive, New York.

Page 109: The Granger Collection, New York.

Page 110: The Bettmann Archive, New York.

Page 114: The Bettmann Archive, New York.

Page 120: Courtesy of Dr. Feynman.

Page 121: Courtesy of Dr. Schwinger.

Page 122: The Granger Collection, New York.

Page 123: The Granger Collection, New York.

Page 124: The Granger Collection, New York.

Page 125: The Granger Collection, New York.

Page 128: The Granger Collection, New York.

Page 142: Courtesy of Dr. Gell-Mann.

Page 179: The Bettmann Archive, New York.